OXFORD LOGIC GUIDES: 12

GENERAL EDITORS
DANA SCOTT
JOHN SHEPHERDSON

OXFORD LOGIC GUIDES

1. Jane Bridge: *Beginning model theory: the completeness theorem and some consequences*
2. Michael Dummett: *Elements of intuitionism*
3. A.S. Troelstra: *Choice sequences: a chapter of intuitionistic mathematics*
4. J.L. Bell: *Boolean-valued models and independence proofs in set theory* (1st edition)
5. Krister Segerberg: *Classical propositional operators: an exercise in the foundations of logic*
6. G.C. Smith: *The Boole–De Morgan correspondence 1842–1864*
7. Alec Fisher: *Formal number theory and computability: a work book*
8. Anand Pillay: *An introduction to stability theory*
9. H.E. Rose: *Subrecursion: functions and hierarchies*
10. Michael Hallett: *Cantorian set theory and limitation of size*
11. Richard Mansfield and Galen Weitkamp: *Recursive aspects of descriptive set theory*
12. J.L. Bell: *Boolean-valued models and independence proofs in set theory* (2nd edition)

BOOLEAN-VALUED MODELS AND INDEPENDENCE PROOFS IN SET THEORY

J. L. Bell

*Reader in Mathematical Logic
London School of Economics*

Second Edition

CLARENDON PRESS · OXFORD
1985

Oxford University Press, Walton Street, Oxford OX2 6DP
London New York Toronto
Delhi Bombay Calcutta Madras Karachi
Kuala Lumpur Singapore Hong Kong Tokyo
Nairobi Dar es Salaam Cape Town
Melbourne Auckland
and associated companies in
Beirut Berlin Ibadan Mexico City Nicosia

Oxford is a trade mark of Oxford University Press

Published in the United States
by Oxford University Press, New York

© J.L. Bell, 1985

All rights reserved. No part of this publication may be reproduced,
stored in a retrieval system, or transmitted, in any form or by any means,
electronic, mechanical, photocopying, recording, or otherwise, without
the prior permission of Oxford University Press

British Library Cataloguing in Publication Data
Bell, J.L.
Bodlean valued models and independence proofs
in set theory.—2nd ed.—(Oxford logic guides; 12)
1. Set theory
I. Title
511.3'22   QA248
ISBN 0-19-853241-5

Library of Congress Cataloging in Publication Data
Bell, J. L. (John Lane)
Boolean-valued models and independence proofs in set theory.
(Oxford logic guides; 12)
1. Axiomatic set theory. 2. Independence (Mathematics)
3. Algebra, Boolean. 4. Model theory.
I. Title. II. Series.
QA248.B44 1985   511.3'2   85-4950
ISBN 0-19-853241-5

Typeset in the United States
Printed in Great Britain
at the University Press, Oxford
by David Stanford
Printer to the University

To Mimi

Foreword
by Dana Scott

Even if we were not required by Russell's Paradox to take care in formulating the axioms of set theory, we would nevertheless have many difficult questions to answer concerning infinite combinatorics and infinite cardinal numbers. Think for a moment of the Axiom of Choice, the Continuum Hypothesis, Souslin's Hypothesis, the questions of the existence of inaccessible and measurable cardinals, the problems of the Lebesgue measurability of projective sets, or of the determinateness of various kinds of infinite games. These are all questions of 'naïve' set theory, many involving little beyond the concept of the arbitrary set of real numbers. Perhaps it is only hindsight (helped on, to be sure, by Gödel's Incompleteness Theorem), but we would have been extraordinarily lucky if these intricate and often fundamental problems were capable of being settled *by logic alone*. By 'logic' here we mean first-order logic (the logic of connectives and quantifiers) together with some rather pure 'ontological' axioms of set existence (like the Comprehension Axiom). An axiom like the Extensionality Axiom, which says that sets are uniquely determined by their elements, is also sufficiently logical in character, because of its almost definitional nature. So we can throw it to the side of logic.

When we come to the Axiom of Choice, we begin to waver: it might be argued that it is implicit in the concept of the totally *arbitrary* set. On the other hand, there could be other notions of what it means to *determine* a set for which it would fail; thus, the act of assuming it is indeed axiomatic: it is 'self-evident' but not just a matter of logic. But then, perhaps it *is* a matter of logic after all, because the *finite* version is provable. In other words, first-order logic is strong enough for some conclusions, but it is in general too weak: we ought perhaps to allow 'infinitary' inferences also. And at this point we begin to wonder what is meant by logic. It would seem rather circular if in making set theory precise, we had to use set theory in order to make logic precise.

With regard to the Continuum Hypothesis most people, I feel, would call this assumption truly axiomatic, since it is so very special in *excluding* certain sets of reals of an intermediate cardinality. True, it can be stated in rather pure terms, and in *second-order* formulations of set theory it would be decided: only we cannot know which way. Thus, the argument for its logical character is rather thin. Certainly Gödel's consistency proof gives us no real evidence. Beautiful as they are, his so-called constructible sets are very special being almost *minimal* in satisfying formal axioms in a first-order language. They just do not capture the notion of set *in general* (and they were not meant to). The constructible universe is extremely interesting in itself (e.g. Jensen showed, among other things, that Souslin's Hypothesis *fails* if $V = L$), but there are very few who would want to assume $V = L$ once and for all. This leaves us more than ever with unsettled feelings as to *where* to draw the line between mathematics and logic.

The events in set theory since 1960 have in some ways made matters even worse. First there was the explosion in large cardinals beginning with the Hanf-Tarski discovery that the first inaccessible (already large enough for a cardinal) was *much* smaller than the first measurable. There was then a most elaborate development in the study of infinite combinatorics by a large number of researchers too numerous to mention here, but among whom Erdös has a central place. On the logical side these results had many model-theoretic consequences and in particular showed that the spectrum of stronger and stronger large-cardinal axioms was *very* finely divided. (The reader can refer to Drake (1974) for a survey.) And the work still goes on. Perhaps this sort of study is not basically disturbing, however, for it just shows—in nearly linear order—that if you want more you have to assume more. It did turn out that the existence of measurable cardinals was *inconsistent* with $V = L$, but so much the worse for the 'unnatural' constructible sets! And some comfort can be gained from the fact that any number of attempts at showing that measurable cardinals do not exist have

Foreword

failed even though much cleverness was expended.

It was in 1963 that we were hit by a real bomb, however, when Paul J. Cohen discovered his method of 'forcing', which started a long chain reaction of independence results stemming from his initial proof of the independence of the Continuum Hypothesis. Set theory could never be the same after Cohen, and there is simply no comparison whatsoever in the sophistication of our knowledge about models for set theory today as contrasted to the pre-Cohen era. One of the most striking consequences of his work is the realization of the extreme *relativity* of the notion of cardinal number. Gödel has shown that, by *cutting down* on the totality of sets, the cardinals (of the model L) would be very well behaved. Cohen showed that by *expanding* the totality of sets the cardinals would be very *ill* behaved. (Tiresomely difficult questions about the possible bad behaviour of *singular* cardinals from model to model is more than sufficient to make the point.) Of course we had realized that the cardinals of L might not be the cardinals of V (indeed, with *fewer* sets there are *more* cardinals, because there is less chance for a one-one correspondence), but we had no idea before Cohen (and those who so quickly jumped into the field after him) how much independence there could be. Thus we can make (if anyone would want to) $2^{\aleph_0} = \aleph_{17}$ and $2^{\aleph_1} = \aleph_{2001}$ in some model or the other, and even with these silly choices the size of $2^{\aleph_2}$ is not at all well determined (except that it has to be greater or equal to $\aleph_{2001}$ and has to avoid certain singular cardinals).

Cohen's achievement lies in being able to *expand* models (countable, standard models) by adding new sets in a very economical fashion: they more or less have only the properties they are *forced* to have by the axioms (or by the truths of the given model). I knew almost all the set-theoreticians of the day, and I think I can say that no one could have guessed that the proof would have gone in just this way. Model-theoretic methods had shown us how many *non-standard* models there were; but Cohen, starting from very primitive first principles, found the way to keep the models *standard* (that is, with a well-ordered collection of ordinals). And moreover his method was very flexible in introducing lots and lots of models—indeed, too many models. Is it not just a bit embarrassing that the currently accepted axioms for set theory (which could be given—as far as they went—

a perfectly natural motivation) simply did *not* determine the concept of infinite set even in the very important region of the continuum?

We should not get the idea that Cohen's method solves all problems. For example, Shoenfield's Absoluteness Lemma shows us why the 'simplest' non-constructible set is Δ^1_3, and thus Cohen's models can only start their independence proofs at that level of the analytic hierarchy. Furthermore, we as yet have no exactly similar model-theoretic independence proofs *from* $V = L$, and this is certainly a very interesting problem.[†] Nevertheless, Cohen's ideas created so many proofs that he himself was convinced that the formalist position in foundations was the rational conclusion. I myself cannot agree, however. I see that there are any number of contradictory set theories, all extending the Zermelo-Fraenkel axioms: but the models are all just models of the first-order axioms, and first-order logic is weak. I still feel that it ought to be possible to have strong axioms which would generate these types of models as submodels of the universe, but where the universe can be thought of as something absolute. Perhaps we would be pushed in the end to say that all sets are *countable* (and that the continuum is not even a set!) when at last all cardinals are absolutely destroyed. But really pleasant axioms have not been produced by me or anyone else, and the suggestion remains speculation. A new idea (or point of view) is needed, and in the meantime all we can do is to study the great *variety* of models. It is the purpose of the present book to give an introduction to this study via the notion of Boolean-valued models. Chapter 2, however, ties up the approach with Cohen's original ideas, though avoiding the technicalities of the ramified languages as is usual in most later presentations.

The idea of using Boolean-valued models to describe forcing was discovered by Solovay in 1965. He was using Borel sets of positive measure as forcing conditions; the complications of seeing just what was true in his model led him, as I remember from a conversation at Stanford around September of that year, to summarize various calculations by saying that

---

[†]Recently, however, Harvey Friedman in his paper "On the necessary use of abstract set theory," Advances in Mathematics, vol. 41(1981), pp. 209–280, showed that interesting combinatorial propositions about Borel functions are independent even from $V = L$. His methods use non-standard models in a way he considers essential. It does not seem that the method can be called similar to Cohen's, however, and there seem to be good reasons for this difference.

Foreword

the combination of conditions forcing a statement added up to the 'value' of that statement (*cf.* Theorem 2.4 in this book and the surrounding discussion). Petr Vopěnka (1965) independently had much the same idea, but his initial presentation was brief and not so very attractive; so we were not much struck by his approach at first. In thinking over Solovay's suggestion, it occurred to me that by *starting* with Boolean-valued sets from the very beginning, many of the more tedious details of Cohen's original construction of the model were avoided. Solovay by November of 1965 had also come to this conclusion himself, and, as it turned out, this was what Vopěnka was actually doing. In the end, as was demonstrated by the paper of Shoenfield (1971) from the 1967 Set Theory Symposium, there is very little to choose between the methods: forcing and Boolean-valued models both come to the same thing. Psychologically, however, one attitude or the other may be more suggestive. Boolean-valued models *are* quite natural; but, when it comes to the proofs (and the *construction* of the right model), one often has to look very closely at the forcing conditions.

The whole history of the independence proofs is rather complex and it could only be made clear by going into the exact technical details. The two volumes of contributions to the 1967 Symposium (Scott (1971) and Jech (1974)) contain many of the original papers by Cohen and an historical paper on the Prague School by Petr Hajek. The lecture notes by Felgner (1971) also contain a very useful summary of basic results with many references to the original sources. In the present book, John Bell has made very good use of the lecture notes on Boolean-valued models which were distributed at the Symposium, and which were prepared for typing by me and Kenneth Bowen during the time of the conference from my handwritten manuscript. These are the notes (Scott (1967)) mentioned in the bibliography; in writing them my main role was that of an expositor.

There are many references in the literature to the Scott-Solovay paper which was to be published as an expanded version of the 1967 notes. This paper, alas, does not exist, and it is my own personal failing for not putting it together from the materials I had at hand. I discussed it several times with Robert Solovay, but we were not at the same institution and could not work very closely together. He drafted parts of certain sections, but he was working on so many papers at the same time that he did not have the opportunity to draft the whole paper. The present book essentially supplants the projected Scott-Solovay paper. Part of my own difficulty about writing the Scott-Solovay paper was the fantastic growth of the field and the speed with which it changed. During the winter of 1968-1969 I became profoundly discouraged because I felt unable to make any original contributions: any ideas I had were either wrong or already known. It is easy enough to say now that I should have been content to be a reporter and expositor, but, at the very moment when one is being left behind, things seem less pleasant. I put these remarks forward not as an excuse but simply as an explanation of why I could not complete what I set out to do.

Looking back on the development of logic and set theory it is very tempting to ask why the independence proofs were not discovered earlier. From the point of view of Boolean algebras we had the needed technical expertise in constructing complete Boolean algebras many years before Cohen. Perhaps model theory up to 1960 had concentrated too much on first-order theories. Actually in September of 1951, in a paper of Alonzo Church delivered at the Mexican Scientific Conference, a suggestion for Boolean-valued models of *type theory* had already been made.[†] However, the suggestion suffered from the fault that not enough care was taken over the Axiom of Extensionality, since Church recommends that at higher types one take *all* functions from one type to the other. Unless the equality relation and membership (application) is treated as on p.14 of the present book, difficulties will arise. These difficulties would have been easily overcome, though, if anyone had tried to develop this clearly stated suggestion.

The 'first-order disease' is most plainly seen in the book by Rasiowa and Sikorski (1963). They had for a number of years considered Boolean-valued models of first-order sentences (as had Tarski, Mostowski, Halmos, and many others working on algebraic logic). Unfortunately they spent most of the time considering *logical validity* (truth in all models) rather than the construction of possibly interesting *particular* models. But even so, with all this machinery, no one thought to ask: how do we interpret second-order quantifiers? It would have been the most natural thing in the

---

[†] The paper was published in English in Boletin de la Sociedad Matematica Mexicana, vol. 10, 1953, pp. 41–52, under the title 'Non-normal truth-tables for the propositional calculus'. The remark Church attributes to Lagerström.

Foreword

logical world, because the values of sentences with *arbitrary* Boolean-valued *relations* were already defined. The step from the arbitrary constant, to the variable, to the quantifier is obvious: it had already been taken at the first-order level. It was a real opportunity missed, and one missed for no good reason except the failure to ask the right question. And, if the question had been asked, the problems about cardinalities would have had to be faced. Well, there is no changing of history.

What about forcing? How new was this idea? As we have said, the application to set theory was strikingly new. Kleene, however, had already used a similar idea in recursion theory where, in studying degrees, he had to force a sequence of Σ_1^0-sentences. The wider model-theoretic significance was not appreciated, though, even if the technique was generalized in recursion theory. In studies of intuitionistic logic (both with Kripke models and with Beth models) the kinds of clauses similar to the forcing definition were quite well known, but it did not seem to occur to anyone to employ intuitionistic logic in making *extensions* of models. Kreisel, in a paper at the Infinitistic Methods Conference in Warsaw in 1959, suggested briefly something very like forcing, but the plan lay quite undeveloped by him or any of his readers. However, after Cohen's original announcement, I pointed out the analogy with intuitionistic interpretations, and along these lines Cohen simplified his treatment of negation at my suggestion. Later the intuitionistic analogy was taken up more seriously by Grzegorczyk and worked out in detail in the book by Fitting (1969). What has transpired since that time is that the set theory in intuitionistic logic proper (not in the double-negative, weak-forcing format) has become much more interesting owing chiefly to the work in category theory by Lawvere and Tierney on topoi. Not only are there Heyting-valued models, but there are many more abstract 'sheaf' models. This is, however, a topic for quite another book, since these new models in intuitionistic logic have not as yet resulted in new independence proofs in *classical* set theory. I think we can look forward to some new insights in this direction, nevertheless, when the more abstract models are better understood.

Oxford, May 1977   (Revised, Pittsburgh, August 1984)

Preface

The present book had its origin in lecture courses I gave at London University during the early seventies. In writing it my objective has been to provide a systematic and adequately motivated account of the theory of Boolean-valued models, deriving along the way the central set-theoretic independence proofs in the particularly elegant form that the Boolean-valued approach enables them to assume. The book is primarily intended for readers who have mastered the material ordinarily dealt with in a first course on axiomatic set theory, including constructible sets and the Gödel relative consistency proofs. I have also assumed some acquaintance with mathematical logic, Boolean algebras, and the rudiments of general topology. In order to expand the scope of the book, and to develop the reader's skill in the subject, many problems (with hints for solution) have been included, some of a more sophisticated character.

The chief purpose in preparing this second edition has been to incorporate an account of iterated Boolean extensions and the consistency (and independence) of Souslin's hypothesis, material not included in the original edition. The addition of this new material (which appears in Chapter 6) necessitated that certain changes be made in earlier chapters: this requirement, together with the generous offer of the Oxford University Press to completely reset the book, afforded me the opportunity of substantially modifying the original text. The major changes in this regard include a new—and I believe more perspicuous—proof that the axiom of choice holds in the model, and a revamping of Chapter 3 on the independence of the axiom of choice in terms of group actions on the model. (I am grateful to Yoshindo Suzuki for suggesting the idea of bringing group actions into the foreground.) I have also grasped the opportunity of suppressing the references with which the original text was liberally peppered, and whose somewhat misleading nature proved vexatious to several reviewers. These references have now been replaced by a set of brief—but nevertheless, I hope, reasonably accurate—historical notes at the end of the book.

It will be evident that in writing a book of this kind I have incurred a heavy intellectual debt to the mathematicians whose work I have attempted—with some temerity, perhaps!—to expound. In this respect I am particularly indebted to Dana Scott. His unpublished, but widely circulated, 1967 notes (Scott 1967) are familiar to all set-theorists as the *urtext* in the field of Boolean-valued models and their influence is to be observed throughout the book. Moreover, as Editor of the Oxford Logic Guides he offered much advice and assistance during the preparation of the original text and has generously provided a Foreword. It was at his suggestion that I embarked on the preparation of this new edition; I am grateful to him both for his encouragement and for his offer to have the text set at Carnegie-Mellon University using the formatting system T_EX (which, as the reader will see, has resulted in a most elegant printed text).

It is also a pleasure to tender my thanks to: John Truss for his careful reading of the manuscript of the original edition and his many ideas for improving it; Enrique Hernandez for his assistance in checking the typescript; Gordon Monro for his valuable comments on Chapter 6, which resulted in the eradication of many errors; Karin Minio and members of the Computer Science Department at Carnegie-Mellon University for their careful handling of the computer preparation of the text; and Mimi Bell and Buffy Fennelly for their expert typing of the manuscript. Finally, I would record my gratitude to Anthony Watkinson and the Oxford University Press, without whom the whole enterprise would never have been brought to fruition.

London, June 1984

Table of Contents

List of Problems   xix

Prerequisites   1

1. Boolean-Valued Models and Their Basic Properties   10

   Construction of the Model   10

   Subalgebras and Their Models   20

   Mixtures and the Maximum Principle   25

   The Truth of the Axioms of Set Theory in $V^{(B)}$ 30

   Ordinals and Constructible Sets in $V^{(B)}$ 38

   Cardinals in $V^{(B)}$ 40

2. Forcing and Some Independence Proofs   48

   The Forcing Relation   48

   Independence of the Axiom of Constructibility and the Continuum Hypothesis   54

   Problems   61

3. Group Actions on $V^{(B)}$ and the Independence of the Axiom of Choice   67

   Group Actions on $V^{(B)}$ 67

The Independence of the Existence of Definable Well-Orderings of $P\omega$	71
The Independence of the Axiom of Choice	75
4. Generic Ultrafilters and Transitive Models of ZFC	**87**
Problems	101
5. Cardinal Collapsing and Some Applications to the Theory of Boolean Algebras	**113**
Cardinal Collapsing	113
Applications to the Theory of Boolean Algebras	117
Problems	120
6. Iterated Boolean Extensions, Martin's Axiom and Souslin's Hypothesis	**121**
Souslin's Hypothesis	121
The Independence of SH	124
Martin's Axiom	128
Iterated Boolean Extensions	131
Further Results on Boolean Algebras	141
The Relative Consistency of SH	147
Problems	152
Historical Notes	**155**
Bibliography	**157**
Index of Symbols	**161**
Index of Terms	**163**

List of Problems

1.24	Σ_1-formulas in $V^{(B)}$	25
1.26	Further properties of mixtures	26
1.29	A variant of the Maximum Principle	28
1.30	The Maximum Principle is equivalent to the axiom of choice	28
1.40	Definite sets	35
1.45	Boolean-valued ordinals	39
1.47	Boolean-valued constructible sets	40
1.53	The κ-chain condition	46
2.4	Boolean completions of non-refined sets	50
2.14	Infinite distributive laws and $V^{(B)}$	61
2.15	Infinite distributive laws and $V^{(B)}$ continued	62
2.16	Weak distributive laws and $V^{(B)}$	62
2.17	κ-closure and $V^{(B)}$	63
2.18	An important set of conditions	64
2.19	Consistency of $2^{\aleph_0} = \aleph_2 + \forall \kappa \geq \aleph_1 [2^\kappa = \kappa^+]$ with ZFC	65
2.20	A further relative consistency result	65
2.21	Consistency of $GCH + P\omega \subseteq L + P\omega_1 \not\subseteq L$ with ZFC	66
3.5	Another characterization of homogeneity	70
3.10	The Boolean-valued subset defined by a formula	72

3.11	Ordinal definable sets in $V^{(B)}$	73
3.12	Complete homomorphisms	73
3.13	Ultrapowers as Boolean extensions	74
4.26	Truth in $M^{(B)}$	101
4.27	Countably M-complete ultrafilters	102
4.28	Atoms in B	102
4.29	Atoms and $M[U]$	102
4.30	A trivial Boolean extension	103
4.31	A transitive model of $\neg AC$	103
4.32	The converse to 4.7 fails	104
4.33	Construction of uncountable transitive models of $ZFC + V \neq L$	104
4.34	Generic sets of conditions	105
4.35	Canonical generic sets and the adjunction of maps	106
4.36	Adjunction of a subset of ω	108
4.37	Intermediate submodels and complete subalgebras	108
4.38	Involutions and generic ultrafilters	110
4.39	The submodel of hereditarily ordinal definable sets	111
5.3	$P\omega \cap L$ can be countable	115
5.4	More on collapsing algebras	115
5.5	Consistency of CH and $\neg$CH with the existence of measurable cardinals	115
5.9	Universal complete Boolean algebras	120
5.10	Homogeneous Boolean algebras	120
6.7	A stronger form of Martin's axiom?	129
6.10	An isomorphism of Boolean algebras	132
6.35	The iteration theorem	152
6.36	More on $\otimes$	153
6.37	The operation inverse to $\otimes$	153
6.38	Injective Boolean algebras	154

Prerequisites

In this book we shall make considerable use of the theory of *Boolean algebras*: to assist the reader we give a short resumé of the basic notions and results of that theory. (For a fuller account, see Halmos 1963, Sikorski 1964 or Bell and Machover 1977.)

A *Boolean algebra* is a structure $\langle B, \vee, \wedge, {}^*, 0_B, 1_B \rangle$ consisting of a set B, two binary operations[†] $\vee, \wedge$ and one unary operation $*$ on B, and two designated elements $0_B, 1_B$ of B satisfying the following conditions: for any $x, y, z \in B$,

$$x \vee y = y \vee x \qquad\qquad x \wedge y = y \wedge x;$$
$$x \vee x = x \qquad\qquad x \wedge x = x;$$
$$(x \vee y) \wedge y = y \qquad\qquad (x \wedge y) \vee y = y;$$
$$(x \vee y) \wedge z = (x \wedge z) \vee (y \wedge z) \qquad (x \wedge y) \vee z = (x \vee z) \wedge (y \vee z);$$
$$x \vee x^* = 1_B \qquad\qquad x \wedge x^* = 0_B;$$
$$0_B \neq 1_B.$$

By convenient abuse of notation, we usually identify a Boolean algebra with its underlying set B and write $0, 1$ for $0_B, 1_B$.

Given a Boolean algebra B, we define the binary relation $\leq$ on B by

$$x \leq y \leftrightarrow x \wedge y = x,$$

or equivalently,

$$x \leq y \leftrightarrow x \vee y = y.$$

[†] Observe that these operations are denoted by the same symbols as those for the logical connectives *disjunction* and *conjunction*, respectively. However, in each case it will be clear from the context which interpretation is to be assigned to a given occurrence of one of these symbols.

The relation $\leq$ is then a partial ordering of B—called the *natural partial ordering*—in which $0_B, 1_B$ are the least and largest elements respectively, and $x \vee y, x \wedge y$ are the supremum and infimum respectively of $\{x, y\}$.

A Boolean algebra may also be described as a partially ordered set $\langle B, \leq \rangle$ which is a distributive lattice with least and largest elements $0_B, 1_B$ such that for each element x there is a unique element x^* for which $x \vee x^* = 1_B, x \wedge x^* = 0_B$.

For $x, y, z \in B$ we write $x \Rightarrow y$ for $x^* \vee y$, $x \Leftrightarrow y$ for $(x \Rightarrow y) \wedge (y \Rightarrow x)$, and $x - y$ for $x \wedge y^*$. We shall frequently employ the rules

$$x \wedge y \leq z \leftrightarrow x \leq (y \Rightarrow z),$$
$$x \Rightarrow (y \Rightarrow z) = (x \wedge y) \Rightarrow z,$$
$$(x \Rightarrow y) = 1 \leftrightarrow x \leq y,$$
$$(x \Leftrightarrow y) = 1 \leftrightarrow x = y,$$
$$y \leq z \rightarrow (x \Rightarrow y) \leq (x \Rightarrow z).$$

A Boolean algebra B is said to be *complete* if, in the natural partial ordering of B, each subset X has a supremum or *join* (written $\bigvee X$ or $\bigvee_B X$) and an infimum or *meet* (written $\bigwedge X$ or $\bigwedge_B X$). If $X = \{x_i : i \in I\}$, or $X = \{x : \phi(x)\}$, we often write $\bigvee_{i \in I} x_i$ or $\bigvee_{\phi(x)} x$ for $\bigvee X$ and $\bigwedge_{i \in I} x_i$ or $\bigwedge_{\phi(x)} x$ for $\bigwedge X$. We shall use the various rules pertaining to $\bigvee$ and $\bigwedge$ freely: e.g.

$$x \vee \bigwedge_{i \in I} x_i = \bigwedge_{i \in I} x \vee x_i,$$
$$x \wedge \bigvee_{i \in I} x_i = \bigvee_{i \in I} x \wedge x_i,$$
$$(\bigvee_{i \in I} x_i)^* = \bigwedge_{i \in I} x_i^*,$$
$$(\bigwedge_{i \in I} x_i)^* = \bigvee_{i \in I} x_i^*,$$
$$\bigwedge_{i \in I} \bigwedge_{j \in J} x_{ij} = \bigwedge_{j \in J} \bigwedge_{i \in I} x_{ij},$$
$$\bigvee_{i \in I} \bigvee_{j \in J} x_{ij} = \bigvee_{j \in J} \bigvee_{i \in I} x_{ij}.$$

The complete Boolean algebras we shall be using in practice are the following:

Prerequisites

(1) *The two-element algebra* $2 = \{0,1\}$ in which $\vee, \wedge, {}^*$ and $\leq$ are defined by: $0 \vee 0 = 0$, $0 \vee 1 = 1$, $1 \vee 1 = 1$; $0 \wedge 0 = 0$, $0 \wedge 1 = 0$, $1 \wedge 1 = 1$; $0^* = 1$, $1^* = 0$, $0 \leq 1$. It is easy to see that 2 is complete.

(2) The *power set algebra* PX *of a set* X. By definition, PX is the family of all subsets of X, and $\vee, \wedge, {}^*$ are just set-theoretic union, intersection, and complementation with respect to X, while $0, 1$ are $\emptyset, X$ respectively. Again, it is easy to verify that PX is complete; in fact for $\{A_i : i \in I\} \subseteq X$ we have

$$\bigvee_{i \in I} A_i = \bigcup_{i \in I} A_i, \quad \bigwedge_{i \in I} A_i = \bigcap_{i \in I} A_i.$$

(3) *The regular open algebra* $\mathrm{RO}(X)$ *of a topological space* X. A subset U of a topological space[†] X is said to be *regular open* if $U = (\overline{U})^\circ$ (where, for any $A \subseteq X$, $\overline{A}$ and A° are, respectively, the closure and the interior of A). One may describe the regular open sets as those open sets which have no "cracks" or "pinholes". The set $\mathrm{RO}(X)$ of all regular open subsets of X is then a complete Boolean algebra—called the *regular open algebra* of X—in which $\leq$ is $\subseteq$, 0 is $\emptyset$, 1 is X, and, for $U, V \in \mathrm{RO}(X)$, $U \vee V = (\overline{U \cup V})^\circ$, $U \wedge V = U \cap V$, and $U^* = X - \overline{U}$. Also, if $\{U_i : i \in I\} \subseteq \mathrm{RO}(X)$,

$$\bigvee_{i \in I} U_i = (\bigcup_{i \in I} U_i)^{-\circ}, \quad \bigwedge_{i \in I} U_i = (\bigcap_{i \in I} U_i)^\circ$$

Remark. Although regular open algebras may seem at first sight somewhat strange objects to study, they in fact play a natural role within the theory of Boolean algebras. Given a Boolean algebra B, it is well-known that B is isomorphic to the algebra of all closed-and-open (clopen) subsets of a suitable topological space X. If we identify B with this algebra, it turns out that $\mathrm{RO}(X)$ is, in a certain sense, the "least" complete Boolean algebra containing B.

A non-empty subset of B is called a *subalgebra* of B if it is closed under the operations $\wedge, \vee, {}^*$ in B. It is clear that each subalgebra of B must contain 0_B and 1_B, also that the subset $\{0_B, 1_B\}$ is a subalgebra of

---

[†]For the (relatively few) notions from general topology we shall use, the reader may consult the first few chapters of Kelley (1955).

B which is a natural copy of the 2-element algebra 2. For this reason we regard 2 as a subalgebra of *every* Boolean algebra.

A *homomorphism* of one Boolean algebra B into another, B', is a map $h: B \to B'$ such that, for all $x, y \in B$, $h(x \wedge y) = h(x) \wedge h(y), h(x \vee y) = h(x) \vee h(y)$, and $h(x^*) = h(x)^*$. It is easy to see that, under these conditions, h must be order-preserving, *i.e.* $x \leq y \to h(x) \leq h(y)$ for any $x, y \in B$. A *homomorphism* $h: B \to B'$ is *complete* if, for any $X \subseteq B$ such that $\bigvee X$ exists in B, $\bigvee h[X]$ exists in B' and equals $h(\bigvee X)$.

An *automorphism* of B is a one-one homomorphism of B onto itself. Note that any automorphism is a complete homomorphism. The collection of all automorphisms of B forms a group under function composition: this group is called the *group of automorphisms* of B.

A *filter* in a Boolean algebra B is a non-empty subset F of B such that

(i) $x \in F, x \leq y \in B \to y \in F$;

(ii) $x, y \in F \to x \wedge y \in F$;

(iii) $0 \notin F$.

A filter F which satisfies, for any $x \in B$,

(iv) $x \in F$ or $x^* \in F$

is called an *ultrafilter*. It is not hard to see that a filter F is an ultrafilter iff it is maximal (under inclusion) in the class of all filters. The basic fact about the existence of ultrafilters is the following (the so-called *Ultrafilter Theorem*): if X is any subset of B such that $x_1 \wedge \ldots \wedge x_n \neq 0$ for any $x_1, \ldots, x_n \in X$, then there is an ultrafilter in B which includes X.

Let S be a family of subsets of a Boolean algebra B, each member X of which has a supremum $\bigvee X$. An ultrafilter U in B is said to be S-*complete* if for all $X \in S$ we have

$$\bigvee X \in U \to X \cap U \neq \emptyset$$

(the reverse implication holding generally). Equivalently, U is S-complete iff, whenever $X \subseteq U$ and $\{x^* : x \in X\} \in S$, then $\bigwedge X \in U$. The important *Rasiowa-Sikorski theorem* states that, if S is *countable*, then given any $x \neq$

Prerequisites

0 in B, there is an S-complete ultrafilter in B which contains x. (Sketch of proof: let $S = \{T_n : n \in \omega\}$ and $t_n = \bigvee T_n$. By induction one can select $b_n \in T_n$ in such a way that $x \wedge (t_0 \Rightarrow b_0) \wedge \ldots \wedge (t_n \Rightarrow b_n) \neq 0$. Any ultrafilter containing $\{x\} \cup \{t_n \Rightarrow b_n : n \in \omega\}$ meets the requirements.)

We shall also assume that the reader is familiar with the development of *axiomatic set theory* up to the consistency of the axiom of choice and the generalized continuum hypothesis. Again, we give a brief review of the notions we shall need. References here include Cohen (1966), Drake (1974), or Bell and Machover (1977).

The *language of set theory* is a first order language $\mathcal{L}$ with equality which also includes a binary predicate symbol $\in$ ('membership'). The *individual* variables $v_o, v_1, \ldots, x, y, z, \ldots$ of $\mathcal{L}$ are understood to range over *sets*, but we shall also allow the formation of *class terms* $\{x : \phi(x)\}$ for each formula $\phi(x)$. The term $\{x : \phi(x)\}$ is understood to mean 'the class of all x such that $\phi(x)$', and a term of this form will simply be called a *(definable) class*. We assume that classes satisfy *Church's scheme*:

$$\forall y [y \in \{x : \phi(x)\} \leftrightarrow \phi(y)].$$

We shall use the following abbreviations:

dom(f)	domain of f
ran(f)	range of f
Px or $P(x)$	power set of x
$f \mid x$	restriction of f to x
$f[X]$	image of X under f
$\langle x, y \rangle$	ordered pair of x, y
x^y	set of all maps of y into x
Fun(f)	'f is a function'
Ord(x)	'x is an ordinal'
L(x)	'x is constructible'.

Also we define the following *classes*:

$$V = \{x : x = x\}$$
$$\mathrm{ORD} = \{x : \mathrm{Ord}(x)\}$$
$$L = \{x : \mathrm{L}(x)\}.$$

V is the *universe of sets*, L the *universe of constructible sets*, and ORD the *class of all ordinals*.

We shall use lower case Greek variables, α, β, γ, ξ, η to range over ordinals.

Zermelo-Fraenkel set theory (ZF) is the theory in L based on the following axioms:

(1) *Extensionality*

$$\forall x \forall y [\forall z (z \in x \leftrightarrow z \in y) \to x = y].$$

(2) *Separation*

$$\forall u \exists v \forall x [x \in v \leftrightarrow x \in u \land \psi(x)]$$

where v is not free in the formula $\psi(x)$.

(3) *Replacement*

$$\forall u [\forall x \in u \exists y \phi(x, y) \to \exists v \forall x \in u \exists y \in v \phi(x, y)]$$

where v is not free in the formula $\phi(x, y)$.

(4) *Union*

$$\forall u \exists v \forall x [x \in v \leftrightarrow \exists y \in u (x \in y)].$$

(5) *Power set*

$$\forall u \exists v \forall x [x \in v \leftrightarrow \forall y \in x (y \in u)].$$

(6) *Infinity*

$$\exists u [\emptyset \in u \land \forall x \in u \exists y \in u (x \in y)].$$

(7) *Regularity*

$$\forall x [\forall y \in x \phi(y) \to \phi(x)] \to \forall x \phi(x)$$

where y is not free in the formula $\phi(x)$.

Prerequisites

The *axiom of choice* (AC) is the sentence

$$\forall x \exists f [\mathrm{Fun}(f) \wedge \mathrm{dom}(f) = x \wedge \forall y \in x [y \neq \emptyset \rightarrow f(y) \in y]].$$

The theory ZF + AC is denoted by ZFC.

We conceive of *ordinals* in such a way that each ordinal is identical with the set of its predecessors. A *cardinal* is an ordinal not equipollent with any smaller ordinal. Customarily we use the Greek letters κ, λ to denote *infinite* cardinals. We assume that the infinite cardinals are enumerated in a sequence $\aleph_0, \aleph_1, \ldots, \aleph_\alpha, \ldots$ or $\omega, \omega_1, \ldots, \omega_\alpha, \ldots$. The next cardinal after κ is written κ^+. Thus if $\kappa = \aleph_\alpha$, then $\kappa^+ = \aleph_{\alpha+1}$. The cardinality of a set x — i.e. the least cardinal equipollent with x — is denoted by $|x|$. When κ and λ are cardinals we frequently write κ^λ for $|\kappa^\lambda|$. (It will be clear from the context whether a given occurrence of κ^λ is intended to mean the collection of maps from λ to κ or the cardinality of this set.) Recall that $|P(x)| = 2^{|x|}$ for any set x. The *continuum hypothesis* (CH) is the statement $2^{\aleph_0} = \aleph_1$. The *generalized continuum hypothesis* (GCH) is the statement $\forall \kappa (2^\kappa = \kappa^+)$. An infinite cardinal κ is said to be *regular* if whenever $|I| < \kappa$ and $\{\lambda_i : i \in I\} \subseteq \kappa$ then $\sum_{i \in I} \lambda_i < \kappa$. Notice that κ^+ is always regular. Also, if GCH holds and λ is regular, then $\lambda^\kappa = \lambda$ for any cardinal $\kappa < \lambda$.

For any infinite cardinal κ we have a *canonical bijection* between κ and $\kappa \times \kappa$. This is obtained as follows. We well-order $\kappa \times \kappa$ by placing $\langle \xi, \eta \rangle$ before $\langle \xi', \eta' \rangle$ provided the ordered triple $\langle \max(\xi, \eta), \xi, \eta \rangle$ lexicographically precedes $\langle \max(\xi', \eta'), \xi', \eta' \rangle$. Then (see Theorem 5.1 of Drake 1964) κ is order isomorphic to $\kappa \times \kappa$ well-ordered in this way and the canonical bijection is the order isomorphism. If for each $\xi < \kappa$ we write $\langle \beta_\xi, \gamma_\xi \rangle$ for the element of $\kappa \times \kappa$ which corresponds to ξ under the canonical bijection, it is not hard to show that $\beta_\xi \leq \xi$.

The *axiom of constructibility* is the sentence $\forall x L(x)$, or, equivalently, $V = L$. We recall the well-known results of Gödel that if ZF is consistent, so is ZF + $V = L$, and that GCH and AC are both provable in the latter theory.

We shall need some facts about *induction* and *recursion* on *well-founded relations*. Let R be a class of ordered pairs. Then R is said to be *well-founded* if for each $x \in V$ the class $\{y : yRx\}$ is a set and for each non-empty set $x \in V$ there is $y \in x$ such that zRy for no $z \in x$. The second condition is equivalent (assuming AC) to the assertion that for no sequence $x_0, x_1, \ldots$ do we have $x_{n+1} R x_n$ for all n.

Given a well-founded relation R, the *principle of induction* on R asserts that if $\phi(x)$ is any formula, then

$$\forall x [\forall y (yRx \to \phi(y)) \to \phi(x)] \to \forall x \phi(x).$$

The *principle of recursion* on R asserts that if F is any class of ordered pairs which defines a single-valued mapping of V into V (such a class is called a *function* on V), then there is a function G on V such that

$$\forall x [G(x) = F(\langle x, G \mid \{y : yRx\}\rangle)].$$

The principles of induction and recursion on well-founded relations are both provable in ZF as schemes.

By the axiom of regularity, the membership relation is well-founded, so we may define the sets V_α by recursion as follows:

$$V_\alpha = \{x : \exists \xi < \alpha [x \subseteq V_\xi]\}.$$

The axiom of regularity implies that $\forall x \exists \alpha (x \in V_\alpha)$, so we may define a function $\mathrm{rank}(x)$ by setting

$$\mathrm{rank}(x) = \text{least } \alpha \text{ such that } x \in V_{\alpha+1}.$$

The relation $\mathrm{rank}(x) < \mathrm{rank}(y)$ is clearly well-founded, so we have the *principle of induction on rank*:

$$\forall x [[\forall y [\mathrm{rank}(y) < \mathrm{rank}(x)] \to \phi(y)] \to \phi(x)] \to \forall x \phi(x).$$

Next, a few remarks on *models* of set theory. An $(\mathcal{L}\text{-})structure$ is a pair $\langle M, E \rangle$ where M is a non-empty set and $E \subseteq M \times M$. *We shall usually identify a given structure with its underlying set M and write 'M' for both.* If E is a well-founded relation, the structure M is said to be *well-founded*. A *transitive $\in$-structure* is a structure of the form $\langle M, \in \mid M \rangle$ in which M is a *transitive set* (*i.e.* satisfies $x \in y \in M \to x \in M$) and $\in \mid M$

is the $\in$-relation restricted to M, i.e. $\in|M = \{\langle x,y\rangle \in M \times M : x \in y\}$. A transitive $\in$-structure which is a model of ZF or ZFC will be called a *transitive $\in$-model* of ZF or ZFC. *Mostowski's collapsing lemma* asserts that if $\langle M, E\rangle$ is a well-founded structure which is a model of the axiom of extensionality, then there is a unique isomorphism h of $\langle M, E\rangle$ onto a transitive $\in$-structure, where h satisfies $h(x) = \{h(y) : yEx\}$ for $x \in M$.

If $\phi(v_1, \ldots, v_n)$ is an $\mathcal{L}$-formula, M is a structure, and $a_1, \ldots, a_n \in M$, we write $M \vDash \phi[a_1, \ldots, a_n]$ for '$a_1, \ldots, a_n$ satisfies $\phi(v_1, \ldots, v_n)$ in M'. If M is a model of ZFC and t is a defined term of ZFC, we let $t^{(M)}$ be the *interpretation* of t in M. This is defined as follows: if t is a *set*, then $t^{(M)}$ is the unique $a \in M$ such that $M \vDash (v_1 = t)[a]$, while if t is a definable class which is *not* a set we put $t^{(M)} = \{a \in M : M \vDash \phi[a]\}$.

An $\mathcal{L}$-formula is said to be *restricted* if every quantifier in it occurs in the form $\forall x \in y$ or $\exists x \in y$ (i.e. $\forall x(x \in y \to \ldots)$ or $\exists x(x \in y \land \ldots)$), or if it can be proved equivalent in ZFC to such a formula. We recall that the formula Ord(x) is restricted. An $\mathcal{L}$-formula is said to be Σ_1 if it can be built up from atomic formulas and their negations using only the logical operations $\land, \lor, \forall x \in y, \exists x$, or if it can be proved equivalent in ZFC to such a formula. We recall that the formula L(x) is Σ_1. Restricted and Σ_1-formulas are important for the following reason. Let $\phi(v_1, \ldots, v_n)$ be an $\mathcal{L}$-formula, let M be a transitive $\in$-model of ZFC and let $a_1, \ldots, a_n \in M$. If ϕ is restricted, then

$$M \vDash \phi[a_1, \ldots, a_n] \leftrightarrow \phi(a_1, \ldots, a_n),$$

while if ϕ is Σ_1, then

$$M \vDash \phi[a_1, \ldots, a_n] \to \phi(a_1, \ldots, a_n).$$

Finally, we point out that, unless otherwise indicated, all theorems, lemmas, problems, etc. in this book are to be regarded as being *proved in* ZFC.

Chapter 1

Boolean-Valued Models and Their Basic Properties

Construction of the Model

Suppose that for each set $x \in V$ we are given a *characteristic function* for x, *i.e.* a function c_x taking values in the Boolean algebra $2 = \{0,1\}$ such that $x \subseteq \text{dom}(c_x)$ and, for all $y \in \text{dom}(c_x)$, $c_x(y) = 1$ iff $y \in x$. It is clear that all information about x is carried by c_x, so it is natural to *identify* x with c_x. If we perform this identification for all $x \in V$, we see that V may, in a natural sense, be regarded as a class of 2-valued functions. The snag here is that, although the process turns each $x \in V$ into a 2-valued function, the function fails to be homogeneous in that its *domain* does *not* (in general) consist of 2-valued functions.

Let us examine this notion of *homogeneity* a little more closely. It is clear that, however we go about defining it, we should require a 2-valued function to be homogeneous iff its domain is a set of homogeneous 2-valued functions. Now this looks very much like a definition by *recursion*; and indeed the recursion in question can be explicitly performed as follows. By transfinite recursion on α we define

$$V^{(2)}_\alpha = \{x : \text{Fun}(x) \land \text{ran}(x) \subseteq 2 \land \exists \xi < \alpha [\text{dom}(x) \subseteq V^{(2)}_\xi]\} \quad (1.1)$$

(compare the definition of the V_α!), and then put

$$V^{(2)} = \{x : \exists \alpha [x \in V^{(2)}_\alpha]\}. \quad (1.2)$$

1. Boolean-Valued Models

$V^{(2)}$ is then the required class of all homogeneous 2-valued functions, since it is easy to see that we have

$$x \in V^{(2)} \leftrightarrow \mathrm{Fun}(x) \wedge \mathrm{ran}(x) \subseteq 2 \wedge \mathrm{dom}(x) \subseteq V^{(2)}. \qquad (1.3)$$

In future we shall drop the cumbersome term 'homogeneous 2-valued function' and call the members of $V^{(2)}$ simply *2-valued sets*. From (1.3) we see that a 2-valued set is a 2-valued function whose domain is a set of 2-valued sets. The class $V^{(2)}$ is called the *universe of 2-valued sets*; we shall see later on that, as expected, it is in a natural sense *isomorphic* to the standard universe V of sets.

What we now propose to do is to replace the Boolean algebra 2 by an arbitrary *complete Boolean algebra* B, thus obtaining what we shall call the *universe* $V^{(B)}$ *of B-valued sets*. We shall show that there is a natural way of assigning to each sentence σ of the language of set theory an element $[\![\sigma]\!]^B$ of B which will act as the 'Boolean truth value' of σ in the universe $V^{(B)}$. Calling the sentence σ *true* in $V^{(B)}$ if $[\![\sigma]\!]^B = 1_B$, and *false* in $V^{(B)}$ if $[\![\sigma]\!]^B = 0_B$ (cf. classical notion of truth and falsehood), we show that, for any complete Boolean algebra B, all the theorems of ZFC are true in $V^{(B)}$, or, to put it more suggestively, that $V^{(B)}$ is a 'Boolean-valued model' of ZFC. On the other hand, we shall see that, by selecting B carefully, we can arrange for a variety of set-theoretic assertions, e.g. the continuum hypothesis or the axiom of constructibility, to be *false* in $V^{(B)}$. The failure of the continuum hypothesis in $V^{(B)}$ will be achieved by selecting B in such a way that, in $V^{(B)}$, ω has many (e.g. $\aleph_2$ or $\aleph_{\omega+1}$) 'B-valued subsets' which are not subsets in the 2-valued sense. In this way we will establish the independence of the continuum hypothesis from ZFC.

We now suppose given a complete Boolean algebra B, which we will assume to be fixed throughout the rest of this chapter. We also assume that B is a *set*, i.e. $B \in V$.

We define the *universe* $V^{(B)}$ *of B-valued sets* by analogy with (1.2); namely, we define, by recursion on α,

$$V_\alpha^{(B)} = \{x : \mathrm{Fun}(x) \wedge \mathrm{ran}(x) \subseteq B \wedge \exists \xi < \alpha [\mathrm{dom}(x) \subseteq V_\xi^{(B)}]\} \qquad (1.4)$$

and

$$V^{(B)} = \{x : \exists \alpha [x \in V_\alpha^{(B)}]\}. \qquad (1.5)$$

We see immediately that, as in (1.3), we have

$$x \in V^{(B)} \leftrightarrow \mathrm{Fun}(x) \land \mathrm{ran}(x) \subseteq B \land \mathrm{dom}(x) \subseteq V^{(B)}, \tag{1.6}$$

i.e. a B-valued set is a B-valued function whose domain is a set of B-valued sets. $V^{(B)}$ is called a *Boolean extension of V*, or, more precisely, the *B-extension of V*.

An easy induction on rank argument proves:

1.7. Induction Principle for $V^{(B)}$. *For any formula $\phi(x)$,*

$$\forall x \in V^{(B)} [\forall y \in \mathrm{dom}(x) \phi(y) \to \phi(x)] \to \forall x \in V^{(B)} \phi(x). \quad \square$$

We now introduce a first-order language suitable for making statements about $V^{(B)}$. Let $\mathcal{L}^{(B)}$ be the first-order language obtained from $\mathcal{L}$ by adding a name for each element of $V^{(B)}$. *For convenience we agree to identify each element of $V^{(B)}$ with its name in $\mathcal{L}^{(B)}$.* By coding the formulas of $\mathcal{L}^{(B)}$ as sets in V in the usual way, it is clear that the collection of formulas of $\mathcal{L}^{(B)}$ becomes a definable class.

At this point it will be convenient to introduce the following terminological convention, which will be adhered to throughout the rest of the book. By *formula*, or *sentence* we shall mean $\mathcal{L}$-formula, or $\mathcal{L}$-sentence, respectively, and by *B-formula*, or *B-sentence* we shall mean $\mathcal{L}^{(B)}$-formula, or $\mathcal{L}^{(B)}$-sentence, respectively.

We next set about constructing the map $[\![\cdot]\!]^B$ from the class of all B-sentences to B which assigns to each B-sentence σ the *Boolean truth value* of σ in $V^{(B)}$.

Suppose for the sake of argument that Boolean truth values have been assigned to all *atomic* B-sentences, *i.e.* sentences of the form $u = v$, $u \in v$, for $u, v \in V^{(B)}$. Then it is natural—by analogy with the classical 2-valued case—to extend the assignment of Boolean truth values to all B-sentences inductively as follows. For B-sentences σ, τ we put

$$[\![\sigma \land \tau]\!]^B =_{df} [\![\sigma]\!]^B \land [\![\tau]\!]^B; \tag{1.8}$$

$$[\![\neg \sigma]\!]^B =_{df} ([\![\sigma]\!]^B)^*. \tag{1.9}$$

1. Boolean-Valued Models

If $\phi(x)$ is a B-formula with one free variable x, such that $[\![\phi(u)]\!]^B$ has been defined for all $u \in V^{(B)}$, we observe that the definable class $\{[\![\phi(u)]\!]^B : u \in V^{(B)}\}$ is a sub*set* of B and define

$$[\![\exists x \phi(x)]\!]^B =_{\mathrm{df}} \bigvee_{u \in V^{(B)}} [\![\phi(u)]\!]^B. \tag{1.10}$$

From (1.8)–(1.10) it follows immediately that

$$[\![\sigma \vee \tau]\!]^B = [\![\sigma]\!]^B \vee [\![\tau]\!]^B; \tag{1.11}$$

$$[\![\sigma \to \tau]\!]^B = [\![\sigma]\!]^B \Rightarrow [\![\tau]\!]^B; \tag{1.12}$$

$$[\![\sigma \leftrightarrow \tau]\!]^B = [\![\sigma]\!]^B \Leftrightarrow [\![\tau]\!]^B; \tag{1.13}$$

$$[\![\forall x \phi(x)]\!]^B = \bigwedge_{u \in V^{(B)}} [\![\phi(u)]\!]^B. \tag{1.14}$$

It remains to assign Boolean truth values to the *atomic B-sentences*. Now we certainly want the *axiom of extensionality* to hold in $V^{(B)}$, so we should have

$$[\![u = v]\!]^B = [\![\forall x \in u[x \in v] \wedge \forall y \in v[y \in u]]\!]^B.$$

Also, in accordance with the *logical* truth $u \in v \leftrightarrow \exists y \in v[u = y]$, which we certainly want to be true in $V^{(B)}$, it should be the case that

$$[\![u \in v]\!]^B = [\![\exists y \in v[u = y]]\!]^B.$$

Finally, we shall require that the Boolean truth value of restricted formulas like $\exists x \in u \phi(x)$ and $\forall x \in u \phi(x)$ depend only on the Boolean truth values of $\phi(x)$ for those x *which are actually in* $\mathrm{dom}(u)$. Moreover, in evaluating the Boolean truth value of such formulas, we agree to be guided by our original case of *characteristic functions*, where, for $x \in \mathrm{dom}(u)$, the 'truth value' of the formula $x \in u$ is $u(x)$. Granted all this, it seems reasonable to require that

$$[\![\exists x \in u \phi(x)]\!]^B = \bigvee_{x \in \mathrm{dom}(u)} [u(x) \wedge [\![\phi(x)]\!]^B]$$

and

$$[\![\forall x \in u \phi(x)]\!]^B = \bigwedge_{x \in \text{dom}(u)} [u(x) \Rightarrow [\![\phi(x)]\!]^B].$$

Putting these things together, we see that we must have, for $u, v \in V^{(B)}$,

$$[\![u \in v]\!]^B = \bigvee_{y \in \text{dom}(v)} [v(y) \wedge [\![u = y]\!]^B]; \qquad (1.15)$$

$$[\![u = v]\!]^B = \bigwedge_{x \in \text{dom}(u)} [u(x) \Rightarrow [\![x \in v]\!]^B] \wedge \\ \bigwedge_{y \in \text{dom}(v)} [v(y) \Rightarrow [\![y \in u]\!]^B]. \qquad (1.16)$$

Now (1.15) and (1.16) may (and shall) be regarded as a *definition* of $[\![u \in v]\!]^B$ and $[\![u = v]\!]^B$ by recursion on a certain well-founded relation. To see this, define for $x, y, u, v \in V^{(B)}$,

$\langle x, y \rangle < \langle u, v \rangle$ iff either $(x \in \text{dom}(u)$ and $y = v)$ or $(x = u$ and $y \in \text{dom}(v))$.

Then $<$ is easily seen to be a well-founded relation on the class $V^{(B)} \times V^{(B)} = \{\langle x, y \rangle : x \in V^{(B)} \wedge y \in V^{(B)}\}$. If we now put, for $u, v \in V^{(B)}$,

$$G(\langle u, v \rangle) = \langle [\![u \in v]\!]^B, [\![v \in u]\!]^B, [\![u = v]\!]^B, [\![v = u]\!]^B \rangle,$$

then (1.15) and (1.16) may be written, for some class function F,

$$G(\langle u, v \rangle) = F(\langle u, v, G \mid \{\langle x, y \rangle : \langle x, y \rangle < \langle u, v \rangle\} \rangle).$$

This constitutes a definition of G by recursion on $<$, and from G we obtain $[\![u \in v]\!]^B$, $[\![u = v]\!]^B$.

Accordingly, we take (1.15) and (1.16) as a *definition* of $[\![\sigma]\!]^B$ for atomic B-sentences σ, and then define $[\![\sigma]\!]^B$ for *all* B-sentences σ by induction on the complexity of σ in accordance with (1.8)–(1.10).

Remarks. 1. The construction of $[\![\sigma]\!]^B$ for arbitrary σ evidently has the form of a *truth definition* for set theory and so cannot be completely formalized within the language of set theory. In fact, although the reader will quickly convince himself that for each specific sentence σ of the language of set theory a specific value for $[\![\sigma]\!]^B$ can be written down within that language, the machinery available in ZFC is not (unless ZFC is inconsistent!) strong enough to formalize the construction of the map $\sigma \mapsto [\![\sigma]\!]^B$ *as a function of* σ. More precisely, one can prove in ZFC that the collection of all pairs $\langle \sigma, [\![\sigma]\!]^B \rangle$ is not a definable class. We must therefore think of this map as being defined *metalinguistically*. This tiresome point can be circumvented by starting with a specific model M of ZFC such that $M \in V$ and performing the whole construction of $V^{(B)}$, $[\![\cdot]\!]^B$ within M; cf. Chapter 4.

2. We observe that there is a considerable *duplication of elements* in $V^{(B)}$. For example, if for each $\alpha \in \mathrm{ORD}$ we define $Z_\alpha \in V^{(B)}$ by $Z_\alpha = \{\langle x, 0_B \rangle : x \in V_\alpha^{(B)}\}$, it is easy to verify that $[\![Z_\alpha = \emptyset]\!]^B = 1$ for all α, so that each of the different members of the proper class $\{Z_\alpha : \alpha \in \mathrm{ORD}\}$ 'represents' the empty set in $V^{(B)}$. In fact it is not hard to show that, for each $u \in V^{(B)}$ there is a *proper class* of $v \in V^{(B)}$ such that $[\![u = v]\!]^B = 1$. Accordingly it is helpful to think of the members of $V^{(B)}$ as 'representatives' or 'labels' for sets[†] (or even 'potential' sets), on which (Boolean-valued) equality is defined as an *equivalence relation* with very large equivalence classes. The duplication of elements in $V^{(B)}$ could be avoided by agreeing to *identify* all elements $u, v \in V^{(B)}$ such that $[\![u = v]\!]^B = 1$, but there would be no particular gain for our purposes.

We say that a B-sentence σ is *true* or *holds with probability* 1 in $V^{(B)}$, and frequently write

$$V^{(B)} \models \sigma,$$

if $[\![\sigma]\!]^B = 1$. A B-formula is *true* in $V^{(B)}$ if its universal closure is true in $V^{(B)}$. Finally, a rule of inference is *valid* in $V^{(B)}$ if it preserves the truth of formulas in $V^{(B)}$.

From now on we shall usually (although not always) take the liberty of dropping the sub- or superscript from $[\![\sigma]\!]^B$, $0_B, 1_B$.

Our next result is basic.

1.17. Theorem. *All the axioms of the first-order predicate calculus with*

[†] $V^{(B)}$ may also be thought of as a 'label space' in the terminology of Cohen (1966).

equality are true in $V^{(B)}$, and all its rules of inference are valid in $V^{(B)}$. In particular, we have

(i) $[\![u = u]\!] = 1$;

(ii) $u(x) \leq [\![x \in u]\!]$ for $x \in \mathrm{dom}(u)$;

(iii) $[\![u = v]\!] = [\![v = u]\!]$;

(iv) $[\![u = v]\!] \wedge [\![v = w]\!] \leq [\![u = w]\!]$;

(v) $[\![u = v]\!] \wedge [\![u \in w]\!] \leq [\![v \in w]\!]$;

(vi) $[\![v = w]\!] \wedge [\![u \in v]\!] \leq [\![u \in w]\!]$;

(vii) $[\![u = v]\!] \wedge [\![\phi(u)]\!] \leq [\![\phi(v)]\!]$,

for any B-formula $\phi(x)$.

Proof. We sketch proofs of (i)–(vii), leaving the rest to the reader.

(i). We employ the induction principle for $V^{(B)}$. Assume as inductive hypothesis that $[\![x = x]\!] = 1$ for $x \in \mathrm{dom}(u)$. Then for $x \in \mathrm{dom}(u)$ we have

(∗) $\quad [\![x \in u]\!] = \bigvee_{y \in \mathrm{dom}(u)} u(y) \wedge [\![x = y]\!] \geq u(x) \wedge [\![x = x]\!] = u(x).$

Therefore $[\![u = u]\!] = \bigwedge_{x \in \mathrm{dom}(u)} [u(x) \Rightarrow [\![x \in u]\!]] = 1$, and the result follows.

(ii) is proved as in (∗) above, using (i).

(iii) holds by symmetry.

(iv) is proved using the induction principle for $V^{(B)}$. Assume as inductive hypothesis that

$$\forall v, w \in V^{(B)} [[\![x = v]\!] \wedge [\![v = w]\!] \leq [\![x = w]\!]]$$

for $x \in \mathrm{dom}(u)$. It follows that, for $x \in \mathrm{dom}(u)$, $y \in \mathrm{dom}(v)$, $z \in \mathrm{dom}(w)$, we have

$$[\![x = y]\!] \wedge [\![y = z]\!] \wedge w(z) \leq [\![x = z]\!] \wedge w(z).$$

1. Boolean-Valued Models

Taking the supremum over z we get, using the definition of $[\![\cdot \in \cdot]\!]$,

$$[\![x = y]\!] \wedge [\![y \in w]\!] \leq [\![x \in w]\!].$$

But from the definition of $[\![\cdot = \cdot]\!]$ we have

$$[\![v = w]\!] \wedge v(y) \leq [\![y \in w]\!]$$

and so

$$[\![v = w]\!] \wedge [\![x = y]\!] \wedge v(y) \leq [\![x \in w]\!].$$

Now take the supremum over y to get

$$[\![x \in v]\!] \wedge [\![v = w]\!] \leq [\![x \in w]\!].$$

Since

$$[\![u = v]\!] \wedge u(x) \leq [\![x \in v]\!],$$

it follows that

$$[\![u = v]\!] \wedge [\![v = w]\!] \wedge u(x) \leq [\![x \in w]\!]$$

or

$$[\![u = v]\!] \wedge [\![v = w]\!] \leq [u(x) \Rightarrow [\![x \in w]\!]].$$

Hence

$$[\![u = v]\!] \wedge [\![v = w]\!] \leq \bigwedge_{x \in \text{dom}(u)} [u(x) \Rightarrow [\![x \in w]\!]], \tag{1}$$

Now, using (iii), the inductive hypothesis implies

$$\forall u, w \in V^{(B)}[[\![w = v]\!] \wedge [\![v = x]\!] \leq [\![w = x]\!]],$$

and, using this, an argument similar to that for (1) yields

$$[\![w = v]\!] \wedge [\![v = u]\!] \leq \bigwedge_{z \in \text{dom}(w)} [w(z) \Rightarrow [\![z \in u]\!]]. \tag{2}$$

Putting (1) and (2) together gives (iv).

(v). If $z \in \text{dom}(w)$, then (iv) gives

$$[\![u = v]\!] \wedge [\![u = z]\!] \wedge w(z) \leq [\![v = z]\!] \wedge w(z).$$

Taking the supremum over z, we get (v).

(vi). If $y \in \text{dom}(v)$, then by definition of $[\![v = w]\!]$ we have

$$[\![v = w]\!] \wedge v(y) \leq [\![y \in w]\!]$$

and so, using (v), we get

$$[\![v = w]\!] \wedge [\![u = y]\!] \wedge v(y) \leq [\![u \in w]\!].$$

Taking the supremum over y gives (vi).

Finally, (vii) is proved by a straightforward induction on the complexity of ϕ, something we leave to the reader. □

Remark. By analogy with the case of characteristic functions, one might expect *equality* to hold in 1.17(ii). Although it is easy to show that this is not the case in general (a task we entrust to the reader), it is nonetheless 'almost' the case. In fact, let us call an element $v \in V^{(B)}$ *extensional* if $v(x) = [\![x \in v]\!]$ whenever $x \in \text{dom}(v)$. Then for each $u \in V^{(B)}$ there is an *extensional* $v \in V^{(B)}$ such that $[\![u = v]\!] = 1$. (Simply put $v = \{\langle x, [\![x \in u]\!]\rangle : x \in \text{dom}(u)\}$.)

It follows from 1.17 that all the *theorems* of first-order predicate calculus are true in $V^{(B)}$.

We can now *prove* the laws governing the assignment of Boolean truth values to formulas with restricted quantifiers.

1.18. Corollary. *For any B-formula $\phi(x)$ with one free variable x, and all $u \in V^{(B)}$,*

(i) $\quad [\![\exists x \in u \phi(x)]\!] = \bigvee_{x \in \text{dom}(u)} [u(x) \wedge [\![\phi(x)]\!]];$

(ii) $\quad [\![\forall x \in u \phi(x)]\!] = \bigwedge_{x \in \text{dom}(u)} [u(x) \Rightarrow [\![\phi(x)]\!]].$

1. Boolean-Valued Models

Proof. We need only establish (i); (ii) then follows by duality. We have

$$
\begin{aligned}
[\![\exists x \in u\, \phi(x)]\!] &= [\![\exists x[x \in u \wedge \phi(x)]]\!] \\
&= \bigvee_{y \in V^{(B)}} [\![y \in u \wedge \phi(y)]\!] \\
&= \bigvee_{y \in V^{(B)}} \bigvee_{x \in \mathrm{dom}(u)} [\![[\![x = y]\!] \wedge u(x) \wedge [\![\phi(y)]\!]]\!] \\
&= \bigvee_{x \in \mathrm{dom}(u)} [u(x) \wedge \bigvee_{y \in V^{(B)}} [\![x = y \wedge \phi(y)]\!]] \\
&= \bigvee_{x \in \mathrm{dom}(u)} [u(x) \wedge [\![\exists y[x = y \wedge \phi(y)]]\!]] \\
&= \bigvee_{x \in \mathrm{dom}(u)} [u(x) \wedge [\![\phi(x)]\!]]. \qquad \square
\end{aligned}
$$

Our next result shows precisely how the properties of $V^{(B)}$ can be used to produce relative consistency proofs in set theory. Given a theory T, write $\mathrm{Consis}(T)$ for 'T is consistent'. Then we have

1.19. Theorem. *Let T, T' be extensions of* ZF *such that* $\mathrm{Consis}(\mathrm{ZF}) \to \mathrm{Consis}(T')$, *and suppose that in $\mathcal{L}$ we can define a constant term B such that:*

(∗) $T' \vdash B$ *is a complete Boolean algebra and, for each axiom τ of T, we have* $T' \vdash [\![\tau]\!]^B = 1_B$.

Then $\mathrm{Consis}(\mathrm{ZF}) \to \mathrm{Consis}(T)$.

Proof. If T is inconsistent, then for some axioms $\tau_1, \ldots, \tau_n$ of T we would have, for any sentence σ,

$$\vdash \tau_1 \wedge \ldots \wedge \tau_n \to \sigma \wedge \neg \sigma. \tag{1}$$

Now let B be a complete Boolean algebra satisfying (∗). Then

$$T' \vdash [\![\tau_1 \wedge \ldots \wedge \tau_n]\!]^B = 1_B. \tag{2}$$

But (1) gives

$$T' \vdash [\![\tau_1 \wedge \ldots \wedge \tau_n]\!]^B \leq [\![\sigma \wedge \neg \sigma]\!]^B = 0_B,$$

so that, by (2)

$$T' \vdash 1_B \leq 0_B,$$

so, T', and hence ZF, would be inconsistent. □

Using the standard techniques of arithmetization, Theorem 1.19 can be stated as follows: if T, T' are extensions of ZF such that (∗) holds and $Consis(\text{ZF}) \to Consis(T')$ is provable in first-order arithmetic, then $Consis(\text{ZF}) \to Consis(T)$ is also provable in first-order arithmetic. Accordingly, 1.19 shows that the method of Boolean-valued models can furnish purely *finitary* relative consistency proofs.

Remark. $V^{(B)}$ is an example of the more general notion of Boolean-valued structure. Given a complete Boolean algebra B, a *B-valued structure* is a triple S consisting of a class S and two maps

$$[\![\cdot = \cdot]\!]_S, [\![\cdot \in \cdot]\!]_S : S \times S \to B$$

satisfying the analogues of (i), (iii)–(vi) of 1.17, *i.e.*,

$$[\![s = s]\!]_S = 1,$$
$$[\![s = t]\!]_S = [\![t = s]\!]_S,$$
$$[\![s = t]\!]_S \wedge [\![t = u]\!]_S \leq [\![s = u]\!]_S,$$
$$[\![s = t]\!]_S \wedge [\![s \in u]\!]_S \leq [\![t \in u]\!]_S,$$
$$[\![t = u]\!]_S \wedge [\![s \in t]\!]_S \leq [\![s \in u]\!]_S,$$

for all $s, t, u \in S$.

The assignment of Boolean values can then be extended recursively to sentences of the language L augmented by names for all elements of S as in 1.8–1.10, *i.e.*,

$$[\![\sigma \wedge \tau]\!]_S = [\![\sigma]\!]_S \wedge [\![\tau]\!]_S,$$
$$[\![\neg \sigma]\!]_S = [\![\sigma]\!]_S^*$$
$$[\![\exists x \phi(x)]\!]_S = \bigvee_{s \in S} [\![\phi(s)]\!]_S,$$

One then shows easily by induction on complexity of sentences that the analogue of 1.17(vii) holds for S, *i.e.* for any L-formula $\phi(x)$ and $s, t \in S$,

$$[\![s = t]\!]_S \wedge [\![\phi(s)]\!]_S \leq [\![\phi(t)]\!]_S.$$

Subalgebras and Their Models

A complete Boolean algebra B' is said to be a *complete subalgebra* of B if B' is a subalgebra of B and, for any $X \subseteq B'$, $\bigvee X$ and $\bigwedge X$ formed in B'

1. Boolean-Valued Models

are the same as $\bigvee X$ and $\bigwedge X$, respectively, formed in B. Our next result shows that, if B' is a complete subalgebra of B, then $V^{(B')}$ is, in a natural sense, a *submodel* of $V^{(B)}$.

1.20. Theorem. *Let B' be a complete subalgebra of B. Then*

(i) $\quad V^{(B')} \subseteq V^{(B)}$.

Moreover, for $u, v \in V^{(B')}$,

(ii) $\quad [\![u \in v]\!]^{B'} = [\![u \in v]\!]^{B}$;

(iii) $\quad [\![u = v]\!]^{B'} = [\![u = v]\!]^{B}$.

Proof. (i) is clear, while (ii) and (iii) are proved simultaneously by induction on the well-founded relation $y \in \text{dom}(x)$. Details are left to the reader. (*Hint*: the inductive hypothesis is: for all $y \in \text{dom}(v)$ and all $u \in V^{(B)}$,

$$[\![u \in y]\!]^{B'} = [\![u \in y]\!]^{B}$$
$$[\![u = y]\!]^{B'} = [\![u = y]\!]^{B}$$
$$[\![y \in u]\!]^{B'} = [\![y \in u]\!]^{B}.) \quad \square$$

1.21. Corollary. *If B' is a complete subalgebra of B, then, for any restricted formula $\phi(v_1, \ldots, v_n)$ and any $u_1, \ldots, u_n \in V^{(B')}$,*

$$[\![\phi(u_1, \ldots, u_n)]\!]^{B'} = [\![\phi(u_1, \ldots, u_n)]\!]^{B}.$$

Proof. By induction on the complexity of ϕ. For atomic ϕ the result holds by 1.20. The only non-trivial induction step arises when ϕ is $\exists x \in u \psi$. And here we argue as follows: if $u, u_1, \ldots, u_n \in V^{(B)}$, then, writing $\bigvee^B$, $\bigvee^{B'}$ for joins in B, B' respectively,

$$[\![\phi(u, u_1, \ldots, u_n)]\!]^{B'} = \bigvee^{B'}_{x \in \text{dom}(u)} [u(x) \wedge [\![\psi(x, u_1, \ldots, u_n)]\!]^{B'}]$$
$$= \bigvee^{B}_{x \in \text{dom}(u)} [u(x) \wedge [\![\psi(x, u_1, \ldots, u_n)]\!]^{B}]$$
$$= [\![\phi(u, u_1, \ldots, u_n)]\!]^{B}. \quad \square$$

In this connection we notice that the 2-element algebra $2 = \{0,1\}$ is a complete subalgebra of *every* complete Boolean algebra B, so that $V^{(2)}$ is a *submodel* of every $V^{(B)}$. We are now going to show that $V^{(2)}$ is, in a certain sense, *isomorphic* to the standard universe V. To this end we make the

1.22. Definition. For each $x \in V$,
$$\hat{x} = \{\langle \hat{y}, 1 \rangle : y \in x\}.$$

This is a definition by recursion on the well-founded relation $y \in x$. Clearly, for each $x \in V$, $\hat{x} \in V^{(2)} \subseteq V^{(B)}$. Also, by 1.20, for $x, y \in V$ we have
$$[\![\hat{x} \in \hat{y}]\!]^B = [\![\hat{x} \in \hat{y}]\!]^2 \in 2$$
$$[\![\hat{x} = \hat{y}]\!] = [\![\hat{x} = \hat{y}]\!]^2 \in 2.$$

We may regard $\hat{x}$ as being the natural 'representative' in $V^{(B)}$ of each $x \in V$, and accordingly members of $V^{(B)}$ of the form $\hat{x}$ are called *standard*. Our next result establishes the main facts about the standard members of $V^{(B)}$.

1.23. Theorem.

(i) For $x \in V$, $u \in V^{(B)}$,
$$[\![u \in \hat{x}]\!] = \bigvee_{y \in x} [\![u = \hat{y}]\!].$$

(ii) For $x, y \in V$,
$$x \in y \leftrightarrow V^{(B)} \models \hat{x} \in \hat{y};$$
$$x = y \leftrightarrow V^{(B)} \models \hat{x} = \hat{y}.$$

(iii) The map $x \mapsto \hat{x}$ is one-one from V into $V^{(2)}$.

(iv) For each $u \in V^{(2)}$ there is a unique $x \in V$ such that $V^{(B)} \models u = \hat{x}$.

1. Boolean-Valued Models

(v) *For any formula $\phi(v_1, \ldots, v_n)$ and any $x_1, \ldots, x_n \in V$,*

$$\phi(x_1, \ldots, x_n) \leftrightarrow V^{(2)} \models \phi(\hat{x}_1, \ldots, \hat{x}_n)$$

and if ϕ is restricted, then

$$\phi(x_1, \ldots, x_n) \leftrightarrow V^{(B)} \models \phi(\hat{x}_1, \ldots, \hat{x}_n).$$

Proof. (i) We have

$$\begin{aligned}
[\![u \in \hat{x}]\!] &= \bigvee_{v \in \mathrm{dom}(\hat{x})} [\![\hat{x}(v) \wedge [\![u = v]\!]]\!] \\
&= \bigvee_{y \in x} [\![\hat{x}(\hat{y}) \wedge [\![u = \hat{y}]\!]]\!] \\
&= \bigvee_{y \in x} [\![u = \hat{y}]\!].
\end{aligned}$$

(ii) is established by induction on $\mathrm{rank}(y)$, the induction hypothesis being: for all z with $\mathrm{rank}(z) < \mathrm{rank}(y)$

$$\begin{aligned}
&\forall x [x \in z \leftrightarrow [\![\hat{x} \in \hat{z}]\!] = 1] \\
&\forall x [x = z \leftrightarrow [\![\hat{x} = \hat{z}]\!] = 1] \\
&\forall x [z \in x \leftrightarrow [\![\hat{z} \in \hat{x}]\!] = 1].
\end{aligned}$$

We leave the tedious but straightforward details to the reader.

(iii) follows immediately from (ii).

(iv) The uniqueness of x follows from (ii). The existence of x is proved by induction on the well-founded relation $x \in \mathrm{dom}(u)$. Suppose then that $u \in V^{(2)}$ and

$$\forall x \in \mathrm{dom}(u) \exists y \in V [[\![x = \hat{y}]\!] = 1].$$

We want to show that, for some $v \in V$, $[\![u = \hat{v}]\!] = 1$. Now

$$[\![u = \hat{v}]\!] = \bigwedge_{x \in \mathrm{dom}(u)} u(x) \Rightarrow [\![x \in \hat{v}]\!] \wedge \bigwedge_{y \in v} [\![\hat{y} \in u]\!].$$

So for $[\![u = \hat{v}]\!] = 1$ it is necessary and sufficient that

$$x \in \mathrm{dom}(u) \rightarrow u(x) \leq [\![x \in \hat{v}]\!] = \bigvee_{y \in v} [\![x = \hat{y}]\!]; \tag{1}$$

$$y \in v \to 1 = [\![\hat{y} \in u]\!] = \bigvee_{x \in \text{dom}(u)} [u(x) \wedge [\![x = \hat{y}]\!]]. \tag{2}$$

Clearly, in order to satisfy (2) we must take

$$v = \{y \in V : \exists x \in \text{dom}(u)[u(x) = 1 \wedge [\![x = \hat{y}]\!] = 1]\}.$$

It follows from (ii) and Replacement that $v \in V$, and an application of the inductive hypothesis shows that v satisfies (1). This proves (iv).

The first part of (v) is proved by induction on the complexity of ϕ, using (ii) and (iv). If ϕ is atomic the result holds by (ii). The only non-trivial induction step arises when ϕ is $\exists x \psi$, and this is handled as follows.

Suppose that $x_1, \ldots, x_n \in V$. If

$$[\![\phi(\hat{x}_1, \ldots, \hat{x}_n)]\!]^2 = 1,$$

then

$$\bigvee_{x \in V^{(2)}} [\![\psi(x, \hat{x}_1, \ldots, \hat{x}_n)]\!]^2 = 1$$

so that

$$[\![\psi(x, \hat{x}_1, \ldots, \hat{x}_n)]\!]^2 = 1$$

for some $x \in V^{(2)}$. But, by (iv), for some $y \in V$ we have $[\![x = \hat{y}]\!]^2 = 1$, so that

$$1 = [\![\psi(x, \hat{x}_1, \ldots, \hat{x}_n)]\!]^2 \wedge [\![x = \hat{y}]\!]^2$$
$$\leq [\![\psi(\hat{y}, \hat{x}_1, \ldots, \hat{x}_n)]\!]^2.$$

The inductive hypothesis now gives $\psi(y, x_1, \ldots, x_n)$, which in turn implies $\phi(x_1, \ldots, x_n)$. The converse is similar.

Finally, the second part of (v) follows from the first part and 1.21. □

Parts (iii), (iv) and (v) of the above theorem show that the universe $V^{(2)}$ of 2-valued sets is, as expected, *isomorphic* to the standard universe V. In particular, it follows from (v) that V and $V^{(2)}$ have the *same true sentences*.

1. Boolean-Valued Models

1.24. Problem. (Σ_1-*formulas in* $V^{(B)}$). Let $\phi(v_1,\ldots,v_n)$ be a Σ_1-formula, and let $x_1,\ldots,x_n \in V$. Show that

$$\phi(x_1,\ldots,x_n) \to V^{(B)} \models \phi(\hat{x}_1,\ldots,\hat{x}_n).$$

Mixtures and the Maximum Principle

We are now going to formulate a useful general method for constructing elements of $V^{(B)}$.

Given a subset $\{a_i : i \in I\} \subseteq B$, and a subset $\{u_i : i \in I\} \subseteq V^{(B)}$, we define the *mixture* $\sum_{i \in I} a_i.u_i$ of $\{u_i : i \in I\}$ *with respect to* $\{a_i : i \in I\}$ to be that element $u \in V^{(B)}$ such that

$$\text{dom}(u) = \bigcup_{i \in I} \text{dom}(u_i)$$

and, for $z \in \text{dom}(u)$,

$$u(z) = \bigvee_{i \in I} [a_i \wedge [\![z \in u_i]\!]].$$

If $I = \{0,1\}$ we write $a_0.u_0 + a_1.u_1$ for $\sum_{i \in I} a_i.u_i$: this is called a *2-term mixture*.

A subset $A \subseteq B$ is called an *antichain* in B if $a \wedge b = 0$ for any distinct elements a, b of A. If an antichain A *is given as an indexed set* $\{a_i : i \in I\}$ *we shall always assume that* $a_i \wedge a_j = 0$ *whenever* $i \neq j$ *in* I. A *partition of unity* in B is an antichain A in B such that $\bigvee A = 1$.

Our next result justifies the use of the term "mixture" by showing that under certain mild conditions (in particular, when $\{a_i : i \in I\}$ is an antichain) $\sum_{i \in I} a_i.u_i$ behaves as if it were obtained by "mixing" the B-valued sets $\{u_i : i \in I\}$ together in (at least) the "proportions" $\{a_i : i \in I\}$.

1.25. Mixing Lemma. *Let* $\{a_i : i \in I\} \subseteq B$, *let* $\{u_i : i \in I\} \subseteq V^{(B)}$ *and put* $\sum_{i \in I} a_i.u_i = u$. *Suppose that, for all* $i, j \in I$,

(∗) $a_i \wedge a_j \leq [\![u_i = u_j]\!].$

Then, for all $i \in I$,

$$a_i \leq [\![u = u_i]\!].$$

In particular, the result holds if $\{a_i : i \in I\}$ is an antichain.

Proof. We have $[\![u = u_i]\!] = a \wedge b$, where

$$a = \bigwedge_{z \in \text{dom}(u)} [u(z) \Rightarrow [\![z \in u_i]\!]]$$
$$b = \bigwedge_{z \in \text{dom}(u_i)} [u_i(z) \Rightarrow [\![z \in u]\!]].$$

If $z \in \text{dom}(u)$, then

$$\begin{aligned}
a_i \wedge u(z) &= \bigvee_{j \in I} a_i \wedge a_j \wedge [\![z \in u_j]\!] \\
&\leq \bigvee_{j \in I} [\![u_i = u_j]\!] \wedge [\![z \in u_j]\!] \qquad \text{(by (*))} \\
&\leq [\![z \in u_i]\!],
\end{aligned}$$

so that $a_i \leq [u(z) \Rightarrow [\![z \in u_i]\!]]$ for any $z \in \text{dom}(u)$, whence $a_i \leq a$. On the other hand, if $z \in \text{dom}(u_i)$, then

$$a_i \wedge u_i(z) \leq a_i \wedge [\![z \in u_i]\!] \leq u(z) \leq [\![z \in u]\!]$$

so that $a_i \leq [u_i(z) \Rightarrow [\![z \in u]\!]]$, whence $a_i \leq b$. Hence $a_i \leq a \wedge b$, and the result follows. □

1.26. Problem. (*Further properties of mixtures*). Let $\{a_i : i \in I\}$ be a partition of unity in B.

(i) Let $\{x_i : i \in I\} \subseteq V$ be such that $x_i \neq x_j$ whenever $i \neq j$. Show that there is $x \in V^{(B)}$ such that $a_i = [\![x = \hat{x}_i]\!]$ for all $i \in I$.

(ii) Let $\{u_i : i \in I\} \subseteq V^{(B)}$ and suppose that $v \in V^{(B)}$ satisfies $a_i \leq [\![v = u_i]\!]$ for all $i \in I$. Show that $V^{(B)} \models v = \sum_{i \in I} a_i \cdot u_i$.

Recall that in (1.10) we assigned a Boolean truth value to the formula $\exists x \phi(x)$ by putting

$$[\![\exists x \phi(x)]\!] = \bigvee_{u \in V^{(B)}} [\![\phi(u)]\!].$$

1. Boolean-Valued Models

We now show, using the Mixing Lemma, that $V^{(B)}$ contains so many members that the supremum on the right side of the above equality is actually *attained* at some element $u \in V^{(B)}$.

1.27. Lemma. (The Maximum Principle). *If $\phi(x)$ is any B-formula, then there is $u \in V^{(B)}$ such that*

$$[\![\exists x \phi(x)]\!] = [\![\phi(u)]\!].$$

In particular, if $V^{(B)} \models \exists x \phi(x)$, then $V^{(B)} \models \phi(u)$ for some $u \in V^{(B)}$.

Proof. By (1.10) we have

$$[\![\exists x \phi(x)]\!] = \bigvee_{u \in V^{(B)}} [\![\phi(u)]\!].$$

Since B is a set, so is $\{[\![\phi(u)]\!] : u \in V^{(B)}\}$ and the axiom of choice implies that there is an ordinal α and a set $\{u_\xi : \xi < \alpha\} \subseteq V^{(B)}$ such that $\{[\![\phi(u)]\!] : u \in V^{(B)}\} = \{[\![\phi(u_\xi)]\!] : \xi < \alpha\}$. Accordingly,

$$[\![\exists x \phi(x)]\!] = \bigvee_{\xi < \alpha} [\![\phi(u_\xi)]\!].$$

For each $\xi < \alpha$, put

$$a_\xi = [\![\phi(u_\xi)]\!] - \bigvee_{\eta < \xi} [\![\phi(u_\eta)]\!].$$

Then $\{a_\xi : \xi < \alpha\}$ is an antichain in B and $a_\xi \leq [\![\phi(u_\xi)]\!]$ for all $\xi < \alpha$. Put $u = \sum_{\xi < \alpha} a_\xi \cdot u_\xi$; then by the Mixing Lemma we have $a_\xi \leq [\![u = u_\xi]\!]$ for all $\xi < \alpha$. Also, clearly,

$$[\![\phi(u)]\!] \leq [\![\exists x \phi(x)]\!].$$

On the other hand,

$$[\![\phi(u)]\!] \geq [\![u = u_\xi]\!] \wedge [\![\phi(u_\xi)]\!] \geq a_\xi$$

so that

$$[\![\phi(u)]\!] \geq \bigvee_{\xi < \alpha} a_\xi = \bigvee_{\xi < \alpha} [\![\phi(u_\xi)]\!] = [\![\exists x \phi(x)]\!]. \quad \square$$

1.28. Corollary. *Let $\phi(x)$ be a B-formula such that $V^{(B)} \models \exists x \phi(x)$.*

(i) *For any $v \in V^{(B)}$ there is a $u \in V^{(B)}$ such that $[\![\phi(u)]\!] = 1$ and $[\![\phi(v)]\!] = [\![u = v]\!]$.*

(ii) *If $\psi(x)$ is a B-formula such that for any $u \in V^{(B)}$, $V^{(B)} \models \phi(u)$ implies $V^{(B)} \models \psi(u)$, then $V^{(B)} \models \forall x [\phi(x) \to \psi(x)]$.*

Proof. (i). Apply the Maximum Principle to obtain $w \in V^{(B)}$ such that $[\![\phi(w)]\!] = 1$, put $b = [\![\phi(v)]\!]$ and $u = b.v + b^*.w$. Then

$$[\![\phi(u)]\!] \geq [\![u = v \wedge \phi(v)]\!] \vee [\![u = w \wedge \phi(w)]\!] \geq b \vee b^* = 1,$$

and $[\![u = v]\!] = [\![u = v]\!] \wedge [\![\phi(u)]\!] \leq [\![\phi(v)]\!]$. Since $[\![u = v]\!] \geq b = [\![\phi(v)]\!]$ by definition of u, the result follows.

(ii). Assume the hypothesis, and let $v \in V^{(B)}$. Using (i), choose $u \in V^{(B)}$ such that $[\![\phi(u)]\!] = 1$ and $[\![\phi(v)]\!] = [\![u = v]\!]$. Then $[\![\psi(u)]\!] = 1$ and

$$[\![\phi(v)]\!] = [\![u = v]\!] = [\![u = v]\!] \wedge [\![\psi(u)]\!] \leq [\![\psi(v)]\!].$$

The result follows. □

1.29. Problem. (*A variant of the Maximum Principle*). *Without* using the axiom of choice, show that, if $V^{(B)} \models \exists! x \phi(x)$, then $V^{(B)} \models \phi(u)$ for some $u \in V^{(B)}$. (Choose a sufficiently large ordinal α such that $1 = [\![\exists x \phi(x)]\!] = \bigvee_{x \in V_\alpha^{(B)}} [\![\phi(x)]\!]$ and define $u \in V^{(B)}$ by $\text{dom}(u) = V_\alpha^{(B)}$, $u(z) = [\![\exists x [\phi(x) \wedge z \in x]]\!]$.)

1.30. Problem. (*The Maximum Principle is equivalent to the axiom of choice*).

(i) Let $\{a_i : i \in I\} \subseteq B$ satisfy $\bigvee_{i \in I} a_i = 1$. A partition of unity $\{b_i : i \in I\}$ in B is called a *disjoint refinement* of $\{a_i : i \in I\}$ if $\forall i \in I [b_i \leq a_i]$. Define $u \in V^{(B)}$ by $\text{dom}(u) = \{\hat{i} : i \in I\}$, $u(\hat{i}) = a_i$ for $i \in I$. Let R be the set of disjoint refinements of $\{a_i : i \in I\}$ and $U = \{v \in V^{(B)} : [\![v \in u]\!] = 1\}$. Show that the map $\{b_i : i \in I\} \mapsto \sum_{i \in I} b_i . \hat{i}$ from R to U is one-one and "onto" U in the sense that, for any $v \in U$ there is a unique $\{b_i : i \in I\} \in R$ such that $[\![\sum_{i \in I} b_i . \hat{i} = v]\!] = 1$.

1. Boolean-Valued Models

(ii) Let Σ_B be the assertion

$$\forall u \in V^{(B)}[\llbracket u \neq \emptyset \rrbracket = 1 \to \exists v \in V^{(B)}[\llbracket v \in u \rrbracket = 1]]$$

("every non-empty B-valued set has an element") and Π_B the assertion: "for any set I, every I-indexed family of elements of B with join 1 has a disjoint refinement." Show without using the axiom of choice that Σ_B and Π_B are equivalent. (Use (i).) Deduce that the assertions "Σ_B holds for every complete Boolean algebra B," and "the Maximum Principle holds in $V^{(B)}$ for every complete Boolean algebra B" are each equivalent to the axiom of choice. (Confine attention to the case in which B is of the form PX for an arbitrary set X.)

We conclude this section by introducing the notion of a core of a Boolean-valued set. Let $u \in V^{(B)}$. A set $v \subseteq V^{(B)}$ is called a *core* for u if the following conditions are satisfied: (i) $\llbracket x \in u \rrbracket = 1$ for all $x \in v$, (ii) for each $y \in V^{(B)}$ such that $\llbracket y \in u \rrbracket = 1$ there is a *unique* $x \in v$ such that $\llbracket x = y \rrbracket = 1$. Thus a core for u represents the class of B-valued objects which are elements of u with probability 1.

1.31. Lemma. *Any $u \in V^{(B)}$ has a core.*

Proof. For each $x \in V^{(B)}$ put

$$f_x = \{\langle z, u(z) \wedge \llbracket z = x \rrbracket \rangle : z \in \mathrm{dom}(u)\}.$$

Using the axiom of replacement we can find a set $w \subseteq V^{(B)}$ such that for each $x \in V^{(B)}$ there is $y \in w$ for which $f_x = f_y$. Now let v be a set obtained by selecting one member from each $\sim$-equivalence class in the set $\{x \in w : \llbracket x \in u \rrbracket = 1\}$, where $\sim$ is defined by $x \sim y \leftrightarrow \llbracket x = y \rrbracket = 1$. It is easily verified that v is a core for u. $\square$

Note that a core of a B-valued set is unique up to bijection in the sense that there is a bijection between any pair of such cores. Observe also that, if u is a B-valued set such that $V^{(B)} \models u \neq \emptyset$, then the Maximum Principle implies that any core of u is non-empty.

The following result, which will prove useful later on, is an immediate consequence of 1.28.

1.32. Lemma. *Suppose that $u \in V^{(B)}$ is such that $V^{(B)} \models u \neq \emptyset$ and* let v be a core for u. Then for any $x \in V^{(B)}$ there is $y \in v$ such that $[\![x = y]\!] = [\![x \in u]\!]$. □

The Truth of the Axioms of Set Theory in $V^{(B)}$

We are now going to show that all of the axioms of ZFC are true in $V^{(B)}$ for any complete Boolean algebra B. (This result is usually expressed by saying that $V^{(B)}$ is a *Boolean valued model of* ZFC.) We do this by *proving* in ZFC that $V^{(B)} \models \sigma$ for each axiom σ of ZFC.

1.33. Theorem. *All the axioms—and hence all the theorems—of* ZFC *are true in* $V^{(B)}$.

We prove this theorem by means of a sequence of lemmas.

1.34. Lemma. *The axiom of extensionality is true in* $V^{(B)}$.

Proof. This follows immediately from (1.16) and 1.18 (ii). □

1.35. Lemma. *The axiom scheme of separation is true in* $V^{(B)}$.

Proof. Recall that the scheme in question is

$$\forall u \exists v \forall x [x \in v \leftrightarrow x \in u \land \psi(x)].$$

To see that each instance is true in $V^{(B)}$, let $u \in V^{(B)}$, define $v \in V^{(B)}$ by $\text{dom}(v) = \text{dom}(u)$ and, for $x \in \text{dom}(v)$,

$$v(x) = u(x) \land [\![\psi(x)]\!].$$

Then

$$[\![\forall x[x \in v \leftrightarrow x \in u \land \psi(x)]]\!]$$
$$= [\![\forall x \in v[x \in u \land \psi(x)]]\!] \land [\![\forall x \in u[\psi(x) \to x \in v]]\!].$$

Now

$$[\![\forall x \in v[x \in u \land \psi(x)]]\!]$$
$$= \bigwedge_{x \in \text{dom}(v)} [[u(x) \land [\![\psi(x)]\!]] \Rightarrow [[\![x \in u]\!] \land [\![\psi(x)]\!]]] = 1,$$

1. Boolean-Valued Models

using 1.17(ii). Similarly,

$$[\![\forall x \in u[\psi(x) \to x \in v]]\!] = 1,$$

and the assertion follows. □

1.36. Lemma. *The axiom scheme of replacement is true in $V^{(B)}$.*

Proof. Recall that this is the scheme

$$\forall u[\forall x \in u \exists y \phi(x,y) \to \exists v \forall x \in u \exists y \in v \phi(x,y)].$$

To show that each instance is true in $V^{(B)}$, notice that, for $u \in V^{(B)}$ we have

$$[\![\forall x \in u \exists y \phi(x,y)]\!] = \bigwedge_{x \in \mathrm{dom}(u)} [u(x) \Rightarrow [\![\exists y \phi(x,y)]\!]]$$
$$= \bigwedge_{x \in \mathrm{dom}(u)} [u(x) \Rightarrow \bigvee_{y \in V^{(B)}} [\![\phi(x,y)]\!]]. \quad (1)$$

Since B is a set, we may invoke the axiom of replacement in V to obtain a map $x \mapsto \alpha_x$ with domain $\mathrm{dom}(u)$ and range a set of ordinals such that, for each $x \in \mathrm{dom}(u)$,

$$\bigvee_{y \in V^{(B)}} [\![\phi(x,y)]\!] = \bigvee_{y \in V^{(B)}_{\alpha_x}} [\![\phi(x,y)]\!]. \quad (2)$$

Let $\alpha = \bigcup \{\alpha_x : x \in \mathrm{dom}(u)\}$. Then, by (2),

$$\bigwedge_{x \in \mathrm{dom}(u)} [u(x) \Rightarrow \bigvee_{y \in V^{(B)}} [\![\phi(x,y)]\!]]$$
$$= \bigwedge_{x \in \mathrm{dom}(u)} [u(x) \Rightarrow \bigvee_{y \in V^{(B)}_{\alpha_x}} [\![\phi(x,y)]\!]] \quad (3)$$
$$\leq \bigwedge_{x \in \mathrm{dom}(u)} [u(x) \Rightarrow \bigvee_{y \in V^{(B)}_{\alpha}} [\![\phi(x,y)]\!]].$$

Now put $v = V^{(B)}_{\alpha} \times \{1\}$; then $v \in V^{(B)}$ and

$$\bigvee_{y \in V^{(B)}_{\alpha}} [\![\phi(x,y)]\!] = [\![\exists y \in v \phi(x,y)]\!].$$

Hence, by (1) and (3),

$$[\![\forall x \in u \exists y \phi(x,y)]\!] \leq \bigwedge_{x \in \mathrm{dom}(u)} [u(x) \Rightarrow [\![\exists y \in v \phi(x,y)]\!]]$$
$$= [\![\forall x \in u \exists y \in v \phi(x,y)]\!].$$

The truth of the axiom scheme of replacement in $V^{(B)}$ follows. $\square$

1.37. Lemma. *The axiom of union is true in $V^{(B)}$.*

Proof. This is the sentence

$$\forall u \exists v \forall x [x \in v \leftrightarrow \exists y \in u [x \in y]].$$

To verify its truth in $V^{(B)}$, let $u \in V^{(B)}$; define $v \in V^{(B)}$ so that $\mathrm{dom}(v) = \bigcup\{\mathrm{dom}(y) : y \in \mathrm{dom}(u)\}$ and

$$v(x) = [\![\exists y \in u[x \in y]]\!]$$

for $x \in \mathrm{dom}(v)$. Then

$$[\![\forall x \in v \exists y \in u[x \in y]]\!]$$
$$= \bigwedge_{x \in \mathrm{dom}(v)} [[\![\exists y \in u[x \in y]]\!] \Rightarrow [\![\exists y \in u[x \in y]]\!]] = 1.$$

Also,

$$[\![\forall y \in u \forall x \in y[x \in v]]\!] = \bigwedge_{y \in \mathrm{dom}(u)} [u(y) \Rightarrow \bigwedge_{x \in \mathrm{dom}(y)} [y(x) \Rightarrow [\![x \in v]\!]]]$$
$$= \bigwedge_{y \in \mathrm{dom}(u)} \bigwedge_{x \in \mathrm{dom}(y)} [u(y) \wedge y(x) \Rightarrow [\![x \in v]\!]]$$
$$= a, \text{ say}.$$

Since $x \in \mathrm{dom}(y)$ and $y \in \mathrm{dom}(u) \to x \in \mathrm{dom}(v)$, we have $[\![x \in v]\!] \geq v(x)$ for $x \in \mathrm{dom}(y)$. Also, for $x \in \mathrm{dom}(y)$ and $y \in \mathrm{dom}(u)$ we have

$$u(y) \wedge y(x) \leq u(y) \wedge [\![x \in y]\!]$$
$$\leq \bigvee_{y \in \mathrm{dom}(u)} [u(y) \wedge [\![x \in y]\!]]$$
$$= [\![\exists y \in u[x \in y]]\!]$$
$$= v(x).$$

1. Boolean-Valued Models

Putting these facts together, we see that

$$a \geq \bigwedge_{y \in \mathrm{dom}(u)} \bigwedge_{x \in \mathrm{dom}(y)} [v(x) \Rightarrow v(x)] = 1$$

and the result follows. □

1.38. Lemma. *The power set axiom is true in $V^{(B)}$.*

Proof. This is the sentence

$$\forall u \exists v \forall x [x \in v \leftrightarrow \forall y \in x [y \in u]].$$

To establish its truth in $V^{(B)}$, let $u \in V^{(B)}$ and define $v \in V^{(B)}$ by

$$\mathrm{dom}(v) = B^{\mathrm{dom}(u)}$$

and, for $x \in \mathrm{dom}(v)$,

$$v(x) = [\![x \subseteq u]\!] = [\![\forall y \in x [y \in u]]\!].$$

It suffices to show that

$$[\![\forall x [x \in v \leftrightarrow x \subseteq u]]\!] = 1.$$

First, we note that

$$\begin{aligned}[] [\![\forall x \in v [x \subseteq v]]\!] &= \bigwedge_{x \in \mathrm{dom}(v)} [v(x) \Rightarrow [\![x \subseteq u]\!]] \\ &= \bigwedge_{x \in \mathrm{dom}(v)} [v(x) \Rightarrow v(x)] \\ &= 1. \end{aligned}$$

It remains to show that

$$[\![\forall x [x \subseteq u \rightarrow x \in v]]\!] = 1. \tag{1}$$

Given $x \in V^{(B)}$, define $x' \in V^{(B)}$ by $\mathrm{dom}(x') = \mathrm{dom}(u)$ and $x'(y) = [\![y \in x]\!]$ for $y \in \mathrm{dom}(x')$. Notice that $x' \in \mathrm{dom}(v)$. We show that

$$[\![x \subseteq u \rightarrow x = x']\!] = 1 \tag{2}$$

and
$$[\![x \subseteq u \to x' \in v]\!] = 1, \tag{3}$$
from which it will follow immediately that
$$[\![x \subseteq u \to x \in v]\!] = 1,$$
which yields (1).

We observe that, for any $y \in V^{(B)}$,
$$[\![y \in x']\!] = \bigvee_{z \in \mathrm{dom}(u)} [x'(z) \wedge [\![z = y]\!]]$$
$$= \bigvee_{z \in \mathrm{dom}(u)} [[\![z \in x]\!] \wedge [\![z = y]\!]] \leq [\![y \in x]\!].$$

Therefore
$$[\![x' \subseteq x]\!] = [\![\forall y[y \in x' \to y \in x]]\!] = 1. \tag{4}$$

Next, for any $y \in V^{(B)}$ we have
$$[\![y \in u \wedge y \in x]\!] = \bigvee_{z \in \mathrm{dom}(u)} [u(z) \wedge [\![y = z]\!] \wedge [\![y \in x]\!]]$$
$$\leq \bigvee_{z \in \mathrm{dom}(u)} [[\![y = z]\!] \wedge [\![z \in x]\!]]$$
$$= \bigvee_{z \in \mathrm{dom}(u)} [[\![y = z]\!] \wedge x'(z)]$$
$$= [\![y \in x']\!],$$
so that $[\![u \cap x \subseteq x']\!] = 1$. Hence, using this and (4), we get
$$[\![x \subseteq u]\!] \leq [\![u \cap x \subseteq x' \wedge x' \subseteq x \wedge x \subseteq u]\!] \leq [\![x = x']\!],$$
which gives (2).

Finally we prove (3). We have
$$[\![x \subseteq u]\!] = [\![\forall y[y \in x \to y \in u]]\!]$$
$$= \bigwedge_{y \in V^{(B)}} [[\![y \in x]\!] \Rightarrow [\![y \in u]\!]]$$
$$\leq \bigwedge_{y \in \mathrm{dom}(x')} [x'(y) \Rightarrow [\![y \in u]\!]]$$
$$= [\![\forall y \in x'[y \in u]]\!]$$
$$= [\![x' \subseteq u]\!] = v(x') \leq [\![x' \in v]\!],$$

1. Boolean-Valued Models   35

since $x' \in \operatorname{dom}(v)$. This immediately gives (3), completing the proof. □

In connection with the proof of 1.38, we make the following

1.39. Definition. For $u \in V^{(B)}$ we define $P^{(B)}(u)$ to be that element $v \in V^{(B)}$ such that $\operatorname{dom}(v) = B^{\operatorname{dom}(u)}$ and, for $x \in \operatorname{dom}(v)$, $v(x) = [\![x \subseteq u]\!]$. $P^{(B)}(u)$ is called the *power set of u in $V^{(B)}$*; the proof of 1.38 shows that it satisfies

$$V^{(B)} \models P^{(B)}(u) = Pu.$$

1.40. Problem. (*Definite sets*). An element $u \in V^{(B)}$ is said to be *definite* if $u(x) = 1$ for all $x \in \operatorname{dom}(u)$. Let $u \in V^{(B)}$ be definite, and define $w \in V^{(B)}$ by $w = B^{\operatorname{dom}(u)} \times \{1\}$. Show that

$$[\![\forall x[x \in w \leftrightarrow x \subseteq u]]\!] = 1.$$

(Use 1.39.) Thus, if u is *definite*, the simple object $B^{\operatorname{dom}(u)} \times \{1\}$ also serves as the power set of u in $V^{(B)}$.

1.41. Lemma. *The axiom of infinity is true in $V^{(B)}$.*

Proof. The axiom in question is the sentence $\exists u \phi(u)$, where $\phi(u)$ is the formula

$$\emptyset \in u \wedge \forall x \in u \exists y \in u(x \in y).$$

Now $\phi(u)$ is obviously a restricted formula, and we certainly have $\phi(\omega)$. Hence, by 1.23(v), we get $[\![\phi(\hat{\omega})]\!] = 1$, and so $[\![\exists u \phi(u)]\!] = 1$. □

1.42. Lemma. *The axiom of regularity is true in $V^{(B)}$.*

Proof. The scheme in question is

$$\forall x[\forall y \in x \phi(y) \to \phi(x)] \to \forall x \phi(x).$$

To see that each instance is true in $V^{(B)}$, first put

$$b = [\![\forall x[\forall y \in x \phi(y) \to \phi(x)]]\!].$$

It now suffices to show that, for any $x \in V^{(B)}$,
$$b \leq [\![\phi(x)]\!].$$

We apply the induction principle for $V^{(B)}$ (1.7). Assume for $y \in \mathrm{dom}(x)$ that $b \leq [\![\phi(y)]\!]$. Then
$$b \leq \bigwedge_{y \in \mathrm{dom}(x)} [\![\phi(y)]\!] \leq \bigwedge_{y \in \mathrm{dom}(x)} [x(y) \Rightarrow [\![\phi(y)]\!]]$$
$$= [\![\forall y \in x\, \phi(y)]\!].$$

But
$$b \leq [\![\forall y \in x\, \phi(y)]\!] \Rightarrow [\![\phi(x)]\!]],$$

so that
$$b \leq [\![\forall y \in x\, \phi(y)]\!] \Rightarrow [\![\phi(x)]\!]] \wedge [\![\forall y \in x\, \phi(y)]\!] \leq [\![\phi(x)]\!],$$

as required. □

In order to verify the *axiom of choice* in $V^{(B)}$ it suffices to verify the set-theoretically equivalent principle *Zorn's lemma*. Recall that a partially ordered set is said to be *inductive* if chains (*i.e.* linearly ordered subsets) in it have upper bounds and Zorn's lemma states that any non-empty inductive partially ordered set has a maximal element.

So finally we prove

1.43. Lemma. *Zorn's lemma, and hence the axiom of choice, is true in* $V^{(B)}$.

Proof. By 1.28(ii), it is enough to show that, for any $x, \leq_X \in V^{(B)}$, if $V^{(B)} \models \langle X, \leq_X \rangle$ *is a non-empty inductive partially ordered set* then $V^{(B)} \models \langle X, \leq_X \rangle$ *has a maximal element*. Suppose then that the antecedent of this implication holds. Let Y be a core for X and define the relation $\leq_Y$ on Y by
$$y \leq_Y y' \leftrightarrow [\![y \leq_X y']\!] = 1$$

for $y, y' \in Y$. It is easy to verify that $\leq_Y$ is a partial ordering on Y; we claim that with this partial ordering Y is inductive. For let C be any chain

1. Boolean-Valued Models

in Y. It is readily shown that $C' = C \times \{1\} \in V^{(B)}$ satisfies

$$V^{(B)} \models C' \text{ is a chain in } X.$$

Accordingly, by the Maximum Principle there is $u \in V^{(B)}$ for which

$$V^{(B)} \models u \text{ is an upper bound for } C' \text{ in } X.$$

Now choose $w \in Y$ such that $[\![w = u]\!] = 1$. Then w is an upper bound for C in Y. For if $x \in C$, then clearly $[\![x \in C']\!] = 1$, whence $[\![x \leq_X u]\!] = 1$ so that $[\![x \leq_X w]\!] = 1$, and $x \leq_Y w$.

Therefore Y is inductive as claimed. By Zorn's lemma in V, Y has a maximal element c. Then $[\![c \in X]\!] = 1$; we claim further that

$$V^{(B)} \models c \text{ is a maximal element of } X. \tag{1}$$

To prove this, take $x \in V^{(B)}$ and apply 1.32 to obtain $y \in Y$ for which $[\![x \in X]\!] = [\![x = y]\!]$. Then

$$[\![c \leq_X x \wedge x \in X]\!] = [\![c \leq_X x \wedge x = y]\!] \leq [\![c \leq_X y]\!]. \tag{2}$$

Now let $v = y.a + c.a^*$, where $a = [\![c \leq_X y]\!]$. Then $[\![v \in X]\!] = 1$ and so there is $z \in Y$ for which $[\![v = z]\!] = 1$. It is easily shown that $[\![c \leq_X v]\!] = 1$, whence $[\![c \leq_X z]\!] = 1$, and so $c \leq_Y z$. Hence $c = z$ by the maximality of c. Therefore

$$\begin{aligned}[][\![c \leq_X y]\!] = a &\leq [\![y = v]\!] \\ &\leq [\![y = v]\!] \wedge [\![v = z]\!] \\ &\leq [\![y = z]\!] = [\![y = c]\!],\end{aligned}$$

and so by (2)

$$\begin{aligned}[][\![c \leq_X x \wedge x \in X]\!] &\leq [\![y = c]\!] \wedge [\![x \in X]\!] \\ &\leq [\![y = c]\!] \wedge [\![x = y]\!] \\ &\leq [\![x = c]\!].\end{aligned}$$

Thus

$$V^{(B)} \models \forall x \in X [c \leq_X x \rightarrow x = c]$$

i.e. (1). This completes the proof. □

The proof of Theorem 1.33 is now complete.

Ordinals and Constructible Sets in $V^{(B)}$

Since the formula $\text{Ord}(x)$ is restricted, it follows from 1.23(v) that $[\![\text{Ord}(\hat{\alpha})]\!] = 1$ for every ordinal α. It is natural to call members of $V^{(B)}$ of the form $\hat{\alpha}$ *standard* ordinals in $V^{(B)}$. Our next result relates the property of being an (arbitrary) ordinal in $V^{(B)}$ to that of being a standard ordinal.

1.44. Theorem. *For all $u \in V^{(B)}$,*
$$[\![\text{Ord}(u)]\!] = \bigvee_{\alpha \in \text{ORD}} [\![u = \hat{\alpha}]\!].$$

Proof. Since $[\![\text{Ord}(\hat{\alpha})]\!] = 1$, we have
$$[\![u = \hat{\alpha}]\!] = [\![u = \hat{\alpha}]\!] \wedge [\![\text{Ord}(\hat{\alpha})]\!] \leq [\![\text{Ord}(u)]\!].$$

Hence
$$\bigvee_{\alpha \in \text{ORD}} [\![u = \hat{\alpha}]\!] \leq [\![\text{Ord}(u)]\!].$$

To establish the reverse inequality, first observe that, since $[\![\hat{\eta} = \hat{\xi}]\!] = 0$ whenever $\eta \neq \xi$ (by 1.23(ii)), the map $\xi \mapsto [\![x = \hat{\xi}]\!]$ is one-one from
$$D_x = \{\xi : [\![x = \hat{\xi}]\!] \neq 0\}$$
into B whenever $x \in \text{dom}(u)$. Since B is a set, so therefore is D_x. Put
$$D = \bigcup_{x \in \text{dom}(u)} D_x.$$

If α_0 is any ordinal greater than every ordinal in D, we have $[\![\hat{\alpha}_0 = x]\!] = 0$ for any $x \in \text{dom}(u)$. Hence
$$[\![\hat{\alpha}_0 \in u]\!] = \bigvee_{x \in \text{dom}(u)} [u(x) \wedge [\![\hat{\alpha}_0 = x]\!]] = 0.$$

1. Boolean-Valued Models

A standard theorem of ZF asserts that $\mathrm{Ord}(u) \wedge \mathrm{Ord}(v) \to u \in v \vee u = v \vee v \in u$; hence

$$[\![\mathrm{Ord}(u)]\!] \leq [\![u \in \hat{\alpha}_0]\!] \vee [\![u = \hat{\alpha}_0]\!] \vee [\![\hat{\alpha}_0 \in u]\!].$$

Since $[\![\hat{\alpha}_0 \in u]\!] = \mathbf{0}$, it follows that

$$[\![\mathrm{Ord}(u)]\!] \leq [\![u \in \hat{\alpha}_0]\!] \vee [\![u = \hat{\alpha}_0]\!] \leq \bigvee_{\alpha \in \mathrm{ORD}} [\![u = \hat{\alpha}]\!]. \quad \square$$

1.45. Problem. (*Boolean-valued ordinals*).

(i) Show that, for any formula $\phi(x)$, $[\![\exists \alpha \phi(\alpha)]\!] = \bigvee_\alpha [\![\phi(\hat{\alpha})]\!]$ and $[\![\forall \alpha \phi(\alpha)]\!] = \bigwedge_\alpha [\![\phi(\hat{\alpha})]\!]$. Thus, quantifications over ordinals in $V^{(B)}$ can be replaced by suprema and infima in B over *standard* ordinals.

(ii) Show that the following conditions on $u \in V^{(B)}$ are equivalent:

(a) $[\![\mathrm{Ord}(u)]\!] = 1$;

(b) there is a set A of ordinals and a partition of unity $\{a_\xi : \xi \in A\}$ in B such that $[\![u = \Sigma_{\xi \in A} a_\xi \cdot \hat{\xi}]\!] = 1$.

Thus the ordinals of $V^{(B)}$ are precisely the mixtures of standard ordinals. Formulate and prove a similar result for the 'natural numbers in $V^{(B)}$'.

The situation for *constructible sets* in $V^{(B)}$ is similar to that for ordinals. In fact, we have

1.46. Theorem. *For all $u \in V^{(B)}$,*

$$[\![L(u)]\!] = \bigvee_{x \in L} [\![u = \hat{x}]\!].$$

Proof. Let L_α be the α-th constructible level. Then, using 1.45, we have

$$[\![L(u)]\!] = [\![\exists \alpha(u \in L_\alpha)]\!] = \bigvee_{\alpha \in \mathrm{ORD}} [\![u \in L_{\hat{\alpha}}]\!].$$

Now the formula $x = L_\alpha$ is Σ_1 (by a well-known result of set theory) so 1.24 gives

$$x = L_\alpha \to [\![\hat{x} = L_{\hat{\alpha}}]\!] = 1,$$

i.e.

$$[\![\hat{L}_\alpha = L_{\hat{\alpha}}]\!] = 1.$$

Therefore

$$\begin{aligned}[\![L(u)]\!] &= \bigvee_{\alpha \in \mathrm{ORD}} [\![u \in L_{\hat{\alpha}}]\!] \\ &= \bigvee_{\alpha \in \mathrm{ORD}} [\![u \in \hat{L}_\alpha]\!] \\ &= \bigvee_{\alpha \in \mathrm{ORD}} \bigvee_{x \in L_\alpha} [\![u = \hat{x}]\!] \\ &= \bigvee_{x \in L} [\![u = \hat{x}]\!],\end{aligned}$$

which is the required result. □

It follows immediately from this theorem that, if $V \neq L$, then $[\![V \neq L]\!] = 1$.

1.47. Problem. (*Boolean-valued constructible sets*). State and prove, for constructible sets, results parallel to those given in 1.45.

Cardinals in $V^{(B)}$

We recall that, for each set x, $|x|$ denotes the *cardinality* of x. Since the formula $|x| = |y|$ is easily seen to be Σ_1, it follows immediately from 1.24 that

$$|x| = |y| \to V^{(B)} \models |\hat{x}| = |\hat{y}|. \tag{1.48}$$

(We shall see later on that the converse does not hold in general.)

1.49. Theorem.

(i) $V^{(B)} \models \hat{\aleph}_0 = \aleph_0$.

1. Boolean-Valued Models

(ii) *For all α,*
$$V^{(B)} \models \hat{\aleph}_\alpha \leq \aleph_{\hat{\alpha}}.$$

Proof. (i). The formula $x = \aleph_0$ is restricted, so the result in question follows easily from 1.23 (v).

(ii) is proved by induction on α. For $\alpha = 0$ the result follows immediately from (i). Suppose now that $\alpha > 0$ and
$$V^{(B)} \models \hat{\aleph}_\beta \leq \aleph_{\hat{\beta}} \qquad (1)$$
for all $\beta < \alpha$. Then we have

$$\begin{aligned}
\aleph_0 \leq \xi < \aleph_\alpha &\to |\xi| = \aleph_\beta \text{ for some } \beta < \alpha \\
&\to V^{(B)} \models |\hat{\xi}| = |\hat{\aleph}_\beta| &&\text{(by 1.48)} \\
&\to V^{(B)} \models |\hat{\xi}| \leq \aleph_{\hat{\beta}} &&\text{(by (1))} \\
&\to V^{(B)} \models |\hat{\xi}| < \aleph_{\hat{\alpha}} \\
&\to V^{(B)} \models \hat{\xi} < \aleph_{\hat{\alpha}}.
\end{aligned}$$

Also,
$$\begin{aligned}
\xi < \aleph_0 &\to V^{(B)} \models \hat{\xi} < \hat{\aleph}_0 \\
&\to V^{(B)} \models \hat{\xi} < \aleph_{\hat{\alpha}}.
\end{aligned}$$

Hence
$$\xi < \aleph_\alpha \to V^{(B)} \models \hat{\xi} < \aleph_{\hat{\alpha}}. \qquad (2)$$

Thus
$$\begin{aligned}
[\![\eta < \hat{\aleph}_\alpha]\!] &= \bigvee_{\xi < \aleph_\alpha} [\![\eta = \hat{\xi}]\!] \\
&= \bigvee_{\xi < \aleph_\alpha} [\![[\![\eta = \hat{\xi}]\!] \wedge [\![\hat{\xi} < \aleph_{\hat{\alpha}}]\!]]\!] &&\text{(by (2))} \\
&\leq [\![\eta < \aleph_{\hat{\alpha}}]\!].
\end{aligned}$$

Therefore
$$V^{(B)} \models \forall \eta [\eta < \hat{\aleph}_\alpha \to \eta < \aleph_{\hat{\alpha}}],$$

whence
$$V^{(B)} \models \hat{\aleph}_\alpha \leq \aleph_{\hat{\alpha}}.$$

This completes the induction step, and the proof. □

Let Card(α) be the formula which asserts that α is a cardinal. Then we have

1.50. Theorem.

(i) $V^{(B)} \models \text{Card}(\hat{\alpha})$ *for any* $\alpha \leq \omega$.

(ii) *If* $V^{(B)} \models \text{Card}(\hat{\alpha})$, *then* $\text{Card}(\alpha)$.

Proof. (i) For $\alpha = \omega$ we already know that $V^{(B)} \models \text{Card}(\hat{\alpha})$ by 1.49 (i). On the other hand, it is a theorem of ZF that $\forall \alpha[\alpha \in \omega \to \text{Card}(\alpha)]$. Hence

$$V^{(B)} \models \forall \alpha[\alpha \in \omega \to \text{Card}(\alpha)].$$

But $V^{(B)} \models \hat{\omega} = \omega$ by 1.49 (i), so that

$$V^{(B)} \models \forall \alpha[\alpha \in \hat{\omega} \to \text{Card}(\alpha)].$$

Hence $\bigwedge_{\alpha \in \omega} [\![\text{Card}(\hat{\alpha})]\!] = 1$, and (i) follows.

(ii) Notice that $\neg \text{Card}(\alpha)$ is a Σ_1-formula, and apply 1.24. □

The formula Card(x) is not Σ_1, so the converse of 1.50(ii) may fail, *i.e.* the property of being a cardinal is not in general preserved under the passage from V to $V^{(B)}$ (cf. Chapter 5). However, there is a simple condition on B which ensures that this property *is* preserved.

A Boolean algebra is said to satisfy the *countable chain condition* (ccc) if every antichain in it is *countable*. (It would seem more reasonable to call this the countable *antichain* condition but the present terminology has the sanction of tradition.)

Complete Boolean algebras satisfying the ccc are readily obtained as follows. Let I be any set and let 2^I have the product topology, where 2 is assigned the discrete topology. Then, as is well-known (see, e.g. Kelley 1955, Prob. 5 $O(f)$), any family of disjoint open sets in 2^I is countable, so *a fortiori* $\text{RO}(2^I)$ satisfies ccc.

We now show that cardinals behave very well in $V^{(B)}$ when B satisfies

1. Boolean-Valued Models

ccc.

1.51. Theorem. *Suppose that B satisfies* ccc. *Then, for any α, and any $x, y \in V$,*

(i) $\mathrm{Card}(\alpha) \to V^{(B)} \models \mathrm{Card}(\hat{\alpha})$;

(ii) $V^{(B)} \models \hat{\aleph}_\alpha = \aleph_{\hat{\alpha}}$;

(iii) $|x| = |y| \leftrightarrow V^{(B)} \models |\hat{x}| = |\hat{y}|$;

(iv) $\mathrm{Card}(\alpha) \wedge \alpha$ *is regular* $\to V^{(B)} \models \hat{\alpha}$ *is regular;*

(v)   *if α is an uncountable regular cardinal, and $\xi \in V^{(B)}$ satisfies $[\![\xi < \hat{\alpha}]\!] = 1$, then there is an ordinal $\beta < \alpha$ such that $[\![\xi < \hat{\beta}]\!] = 1$.*

Proof. (i). Let α be a cardinal. If $\alpha \leq \omega$ then $V^{(B)} \models \mathrm{Card}(\hat{\alpha})$ by 1.50(i), so we may suppose that $\alpha > \omega$. To obtain the required conclusion it suffices to show that, for all $f \in V^{(B)}$ and all $\beta < \alpha$,

$$[\![\mathrm{Fun}(f) \wedge \mathrm{dom}(f) = \hat{\beta} \wedge \mathrm{ran}(f) = \hat{\alpha}]\!] = 0.$$

Suppose on the contrary that, for some $f \in V^{(B)}$ and $\beta < \alpha$ we have

$$a = [\![\mathrm{Fun}(f) \wedge \mathrm{dom}(f) = \hat{\beta} \wedge \mathrm{ran}(f) = \hat{\alpha}]\!] \neq 0;$$

then

$$0 \neq a \leq \bigwedge_{\eta < \alpha} \bigvee_{\xi < \beta} [\![f(\hat{\xi}) = \hat{\eta}]\!] \wedge a.$$

It follows that for each $\eta < \alpha$ there is a *least* $\xi_\eta < \beta$ such that

$$[\![f(\hat{\xi}_\eta) = \hat{\eta}]\!] \wedge a \neq 0.$$

Since α is an uncountable cardinal and $\beta < \alpha$ there must exist a $\gamma < \beta$ such that the set

$$X = \{\eta < \alpha : \xi_\eta = \gamma\}$$

is uncountable. It follows immediately that the set

$$\{[\![f(\hat{\gamma}) = \hat{\eta}]\!] \wedge a : \eta \in X\}$$

is an uncountable antichain in B, contradicting ccc. Hence $a = 0$ and (i) follows.

(ii). This goes by induction on α. Assuming that $V^{(B)} \models \hat{\aleph}_\beta = \aleph_{\hat{\beta}}$ for all $\beta < \alpha$, by 1.49(ii) it suffices to show that
$$V^{(B)} \models \aleph_{\hat{\alpha}} \leq \hat{\aleph}_\alpha.$$

By (i), we have $V^{(B)} \models \operatorname{Card}(\hat{\aleph}_\alpha)$. Also, if $\beta < \alpha$, then $V^{(B)} \models \hat{\aleph}_\beta < \hat{\aleph}_\alpha$ and by inductive hypothesis $V^{(B)} \models \hat{\aleph}_\beta = \aleph_{\hat{\beta}}$. Hence $V^{(B)} \models \aleph_{\hat{\beta}} < \hat{\aleph}_\alpha$, so that
$$1 = [\![\operatorname{Card}(\hat{\aleph}_{\hat{\alpha}})]\!] \wedge \bigwedge_{\beta < \alpha} [\![\aleph_{\hat{\beta}} < \hat{\aleph}_\alpha]\!]$$
$$= [\![\operatorname{Card}(\hat{\aleph}_\alpha) \wedge \forall \beta < \hat{\alpha}(\aleph_\beta < \hat{\aleph}_\alpha)]\!]$$
$$\leq [\![\aleph_{\hat{\alpha}} \leq \hat{\aleph}_\alpha]\!],$$

completing the induction step, and proving (ii).

(iii) is an immediate consequence of (ii).

(iv). Let α be a regular cardinal; without loss of generality we may assume $\alpha > \aleph_0$. Suppose that the conclusion is false, i.e., $[\![\hat{\alpha} \text{ is singular}]\!] \neq 0$. Let $\phi(x, y)$ be the statement $\operatorname{Fun}(x) \wedge \operatorname{dom}(x) = y$ and $\operatorname{ran}(x)$ is cofinal in $\hat{\alpha}$. Then
$$0 \neq [\![\hat{\alpha} \text{ is singular}]\!] = [\![\exists \xi < \hat{\alpha} \exists f \phi(f, \xi)]\!] = \bigvee_{\beta < \alpha} [\![\exists f \phi(f, \hat{\beta})]\!].$$

Hence there is $\beta < \alpha$ such that
$$0 \neq [\![\exists f \phi(f, \hat{\beta})]\!] = a, \text{ say},$$

and so the Maximum Principle yields an $f \in V^{(B)}$ for which $a = [\![\phi(f, \beta)]\!]$. Then
$$0 \neq a \leq [\![\operatorname{ran}(f) \text{ is cofinal in } \hat{\alpha}]\!]$$
$$= \bigwedge_{\eta < \alpha} \bigvee_{\xi < \beta} \bigvee_{\mu \geq \eta} [\![f(\hat{\xi}) = \hat{\mu}]\!].$$

It follows that for each $\eta < \alpha$ there are ordinals $\xi_\eta < \beta, \mu_\eta < \alpha$ such that $[\![f(\hat{\xi}_\eta) = \hat{\mu}_\eta]\!] \wedge a \neq 0$. Since α is regular and $\beta < \alpha$ there is $\gamma < \beta$ such

1. Boolean-Valued Models 45

that $X = \{\eta < \alpha : \xi_\eta = \gamma\}$ has cardinality α. Then
$$\{[\![f(\hat{\gamma}) = \hat{\mu}_\eta]\!] \wedge a : \eta \in X\}$$
is an antichain of cardinality $\alpha > \aleph_0$ in B, contradicting ccc. (iv) follows.

(v). Assuming the hypotheses, the set $\{[\![\xi = \hat{\eta}]\!] : \eta < \alpha\}$ is an antichain in B and hence the set $X = \{\eta < \alpha : [\![\xi = \hat{\eta}]\!] \neq 0\}$ is countable. Put $\beta = \sup X + 1$; then $\beta < \alpha$ and we have
$$1 = [\![\xi < \hat{\alpha}]\!] = \bigvee_{\eta < \alpha} [\![\xi = \hat{\eta}]\!] = \bigvee_{\eta \in X} [\![\xi = \hat{\eta}]\!]$$
$$\leq \bigvee_{\eta < \beta} [\![\xi = \hat{\eta}]\!] = [\![\xi < \hat{\beta}]\!]$$
and (v) follows. □

Finally, we prove a result which will be useful in estimating cardinalities in $V^{(B)}$. We shall need the notion of *ordered pair* in $V^{(B)}$: for $u, v \in V^{(B)}$ we define
$$\{u\}^{(B)} = \{\langle u, 1 \rangle\}$$
$$\{u, v\}^{(B)} = \{u\}^{(B)} \cup \{v\}^{(B)}$$
$$\langle u, v \rangle^{(B)} = \{\{u\}^{(B)}, \{u, v\}^{(B)}\}^{(B)}.$$
It is then easily verified that
$$V^{(B)} \models \forall x \forall y \forall u \forall v [\langle x, y \rangle^{(B)} = \langle u, v \rangle^{(B)} \leftrightarrow x = u \wedge y = v].$$

1.52. Lemma. *For any $u \in V^{(B)}$ we can find $f \in V^{(B)}$ such that*
$$V^{(B)} \models \mathrm{Fun}(f) \wedge \mathrm{dom}(f) = \mathrm{dom}(u)\hat{\ } \wedge u \subseteq \mathrm{ran}(f)$$
and hence
$$V^{(B)} \models |u| \leq |\mathrm{dom}(u)\hat{\ }|.$$

Proof. Define
$$f = \{\langle \hat{z}, z \rangle^{(B)} : z \in \mathrm{dom}(u)\} \times \{1\}.$$

Then it is easily verified that $f \in V^{(B)}$ meets the required conditions; to indicate the idea of the proof we show e.g. that $[\![u \subseteq \mathrm{ran}(f)]\!] = 1$. For we have

$$\begin{aligned}
[\![\exists x.\langle x,y\rangle \in f]\!] &= \bigvee_{x \in V^{(B)}} [\![\langle x,y\rangle \in f]\!] \\
&= \bigvee_{z \in \mathrm{dom}(u)} [\![y=z]\!] \wedge \bigvee_{x \in V^{(B)}} [\![x = \hat{z}]\!] \\
&= \bigvee_{z \in \mathrm{dom}(u)} [\![y=z]\!] \\
&\geq \bigvee_{z \in \mathrm{dom}(u)} u(z) \wedge [\![y=z]\!] \\
&= [\![y \in u]\!].
\end{aligned}$$

The other conditions are verified similarly. □

Note that 1.52 yields an alternative proof that the axiom of choice holds in $V^{(B)}$. For, given $u \in V^{(B)}$, there is by the well-ordering theorem in V an ordinal α and a bijection g of α onto $\mathrm{dom}(u)$. It follows that

$$V^{(B)} \models \hat{g} \text{ is a bijection of the ordinal } \hat{\alpha} \text{ onto } \mathrm{dom}(u)\hat{\ }.$$

If $f \in V^{(B)}$ is as specified in 1.52, then

$$V^{(B)} \models f \circ \hat{g} \text{ is a function with domain } \hat{\alpha} \text{ and range } \supseteq u$$

and so

$$V^{(B)} \models u \text{ is well-orderable }.$$

Since this holds for arbitrary $u \in V^{(B)}$, $V^{(B)} \models \mathrm{AC}$.

1.53. Problem. (*The κ-chain condition*). Let κ be an infinite cardinal. B is said to satisfy the *κ-chain condition* (κ-cc) if each antichain in B has cardinality $< \kappa$. (Thus the ccc is the $\aleph_1$-cc.)

(i) Show that B always satisfies $|B|$-cc. (If B contains an antichain of cardinality κ, then $2^\kappa \leq |B|$.)

Now assume that B satisfies κ-cc. Show that:

(ii) $V^{(B)} \models \mathrm{Card}(\hat{\alpha})$ for any cardinal $\alpha > \kappa$;

(iii) if κ is regular, $V^{(B)} \models \mathrm{Card}(\hat{\kappa})$;

1. Boolean-Valued Models

(iv) for each $X \subseteq B$ there is $Y \subseteq X$ such that $|Y| < \kappa$ and $\bigvee X = \bigvee Y$. (Let A be a maximal antichain in the ideal generated by X; show that $\bigvee X = \bigvee A$. For each $a \in A$ there is a finite subset $F_a \subseteq X$ such that $a \leq \bigvee F_a$; show that $Y = \bigcup_{a \in A} F_a$ meets the requirements.)

Chapter 2

Forcing and Some Independence Proofs

The Forcing Relation

Let $P = \langle P, \leq \rangle$ be a fixed but arbitrary partially ordered set. (We shall use letters p, q, r, p', q', r' to denote elements of P.) Intuitively, the elements of P are to be thought of as states of information about or *conditions* on a set-theoretic state of affairs and the relation $p \leq q$ is to be understood as asserting 'p *refines* q' or 'the information content of p includes that of q'. Two elements p and q of P are said to be *compatible*—written $\text{Comp}(p, q)$—if there is $r \in P$ such that $r \leq p$ and $r \leq q$. The relation $\text{Comp}(p, q)$ is intended to express the assertion that p and q are mutually consistent conditions. P is said to be *refined* if

$$\forall p, q \in P [q \not\leq p \to \exists p' \leq q \neg \text{Comp}(p, p')].$$

Thus P is refined if whenever q is not a refinement of p, q has a refinement which is incompatible with p.

For each $p \in P$, put

$$O_p = \{q \in P : q \leq p\}.$$

Then, as is easily verified, the O_p form a base for a topology on P called the (left) *order topology*. We put $\text{RO}(P)$ for the complete Boolean algebra of regular open sets in this topology.

Let us call a subset X of a Boolean algebra B *dense* if $0 \notin X$ and for each $0 \neq b \in B$ there is $x \in X$ such that $x \leq b$.

2. Forcing and Independence Proofs

2.1. Lemma.

(i) P is refined if and only if $O_p \in \mathrm{RO}(P)$ for all $p \in P$.

(ii) If P is refined, the map $p \mapsto O_p$ is an order-isomorphism of P onto a dense subset of $\mathrm{RO}(P)$.

Proof. (i) It is easily verified that, if P is assigned the order topology, then the interior of the closure of a subset X of P is

$$(\overline{X})^\circ = \{q \in P : \forall p' \leq q \exists r \in X [r \leq p']\}.$$

Hence

$$\begin{aligned}(\overline{O_p})^\circ &= \{q \in P : \forall p' \leq q \exists r \leq p [r \leq p']\} \\ &= \{q \in P : \forall p' \leq q \,\mathrm{Comp}(p,p')\}.\end{aligned} \quad (1)$$

Now suppose that P is refined. We automatically have $O_p \subseteq (\overline{O_p})^\circ$ since O_p is open. Conversely, if $q \notin O_p$, then $q \not\leq p$, so since P is refined, there is $p' \leq q$ such that $\neg\mathrm{Comp}(p,p')$ and it follows from (1) that $q \notin (\overline{O_p})^\circ$. Therefore $O_p = (\overline{O_p})^\circ$, i.e. $O_p \in \mathrm{RO}(P)$. Conversely, if $O_p \in \mathrm{RO}(P)$, then $O_p = (\overline{O_p})^\circ$, so, using (1), we have $q \not\leq p \to q \notin O_p \to q \notin (\overline{O_p})^\circ \to \exists p' \leq q \neg \mathrm{Comp}(p,p')$, so P is refined. This proves (i).

(ii) follows easily from (i) and the definition of the order topology on P. □

2.2. Corollary. *P is refined iff it is order-isomorphic to a dense subset of a complete Boolean algebra.*

Proof. Necessity follows from 2.1(ii). Conversely, suppose that P is order-isomorphic to a dense subset of a (complete) Boolean algebra B. Then we may identify P with a dense subset of B. If $p,q \in P$ and $q \not\leq p$, then (in B) $q \wedge p^* \neq 0$, so since P is dense there is $p' \in P$ such that $p' \leq q \wedge p^*$. Thus $p' \leq q$ and it is easy to verify that $\neg\mathrm{Comp}(p,p')$. Therefore P is refined. □

Let us say that a pair $\langle B, e \rangle$ (or simply B) is a *Boolean completion* of P if the following conditions are met:

(i) B is a complete Boolean algebra;

(ii) e is an order-isomorphism of P onto a dense subset of B.

2.3. Lemma. *If $\langle B, e\rangle$ and $\langle B', e'\rangle$ are Boolean completions of P, then there is an isomorphism between B and B' which interchanges $e[P]$ and $e[P']$.*

Proof. We give a sketch, leaving the reader to fill in the details. For each $x \in B$ put $P_x = \{p \in P : e(p) \leq x\}$. Then the density of $e[P]$ in B implies that $x = \bigvee e[P_x]$ for each $x \in B$. The map $f : B \to B'$ defined for $x \in B$ by $f(x) = \bigvee e'[P_x]$ then meets the requirements. □

2.2 and 2.3 imply that *each refined partially ordered set P has a Boolean completion which is unique up to isomorphism.*

If P is refined and $\langle B, e\rangle$ is a Boolean completion of P, P will be called a *basis* or *set of conditions* for B (with respect to e). Under these conditions, we shall frequently *identify* P with its image $e[P]$ in B, so that *P will be regarded as a dense subset of B.*

Remark. The notion of the Boolean completion of a partially ordered set is closely related to the—possibly more familiar—notion of the completion of a Boolean algebra. A (minimal) completion of a Boolean algebra A is a pair $\langle B, f\rangle$ in which B is a complete Boolean algebra and f is a complete monomorphism of A into B such that $f[A - \{0\}]$ is dense in B.

One easily shows (as in Lemma 2.3) that if $\langle B, f\rangle$ and $\langle B', f'\rangle$ are completions of A, then there is an isomorphism between B and B' which interchanges $f[A]$ and $f'[A]$. We can obtain the completion $\langle B, f\rangle$ of A in either of the following two equivalent ways: (1) take $\langle B, e\rangle$ to be a Boolean completion of the partially ordered set $A - \{0\}$ and define $f : A \to B$ by $f(x) = e(x)$ if $x \neq 0$ and $f(0) = 0$; (2) take B to be the regular open algebra of the Stone space of A and f the natural monomorphism of A into B. The completion $\langle B, f\rangle$ of A is characterized by the following universal property: for any complete Boolean algebra C and any complete homomorphism $g : A \to C$, there is a unique complete homomorphism $h : B \to C$ such that $g = h \circ f$.

2.4. Problem. (*Boolean completions of non-refined sets*). Let $\langle P, \leq\rangle$ be

2. Forcing and Independence Proofs

a partially ordered set.

(i) Show that there is a refined partially ordered set $\langle Q, \preceq \rangle$ and an order preserving map j of P onto Q such that, for any $p, q \in P$, $\text{Comp}(p, q) \leftrightarrow \text{Comp}(jp, jq)$. (Define the equivalence relation $\sim$ on P by $p \sim q \leftrightarrow \forall x[\text{Comp}(p, x) \leftrightarrow \text{Comp}(q, x)]$, and take $Q = P/\sim$.)

(ii) Show that $\langle Q, \preceq \rangle$ is uniquely determined up to isomorphism.

The partially ordered set Q is called the *refined associate* of P and the map j is called the *canonical* map. If P is refined, we may take $Q = P$ and j to be the identity.

(iii) Let B be the Boolean completion of Q; then Q may be identified with a dense subset of B and the canonical map j may be regarded as carrying P into B. Show that j is an order-preserving map onto a dense subset of B such that, for $p, q \in P$, $\text{Comp}(p, q) \leftrightarrow j(p) \wedge j(q) \neq 0$.

The algebra B is called the *Boolean completion* of P.

Now let x and y be non-empty sets, where y has at least two elements. We put $C(x, y)$ for the set of all mappings with domain a *finite* subset of x and range a subset of y. We agree to partially order $C(x, y)$ by $\supseteq$, i.e. *inverse* inclusion, and it is easy to verify that this turns $C(x, y)$ into a *refined* partially ordered set. For $p \in C(x, y)$ we put $N(p)$

$$N(p) = \{f \in y^x : p \subseteq f\}$$

where y^x is, as usual, the set of all mappings from x into y. Subsets of y^x of the form $N(p)$ form a base for the *product topology* on y^x, when y is assigned the discrete topology. Each $N(p)$ is then a *clopen* (*i.e.* closed-and-open) set in this topology. Thus, in particular, each $N(p)$ is a *regular open* subset of y^x, and it is easy to verify that the map $p \mapsto N(p)$ is an order-isomorphism of $C(x, y)$ onto a dense subset of $\text{RO}(y^x)$. Therefore $\langle \text{RO}(y^x), N \rangle$ is a *Boolean completion* of $C(x, y)$, and the latter is (up to isomorphism) a *basis* for $\text{RO}(y^x)$.

Remark. In Cohen's original development of forcing, the ordering of the set of forcing conditions P 'goes up', rather than 'down' as we have taken it. In other words, Cohen holds $p \leq q$ to mean that q contains *more* information than p. In particular, the set $P = C(x,y)$ would be taken to be ordered by *inclusion* and not, as we have stipulated, by inverse inclusion. This (the "more means more" convention) is of course eminently reasonable, but unfortunately P would then, in a natural sense, be *anti*-refined rather than refined, and one could only show that P is order *anti*-isomorphic to a dense subset (or, equivalently, isomorphic to an *anti*-dense subset) of a complete Boolean algebra. (To obtain the 'anti'-version of an order-theoretic concept, interchange '$\leq$' and '$\geq$' and '0' and '1'.) Since anti-isomorphisms and anti-dense subsets are not very convenient technically, we have chosen to reverse the more usual ordering of P and thereby adopt the "more means less" convention, thus enabling us to use the more familiar machinery of isomorphisms and dense subsets.

Now let B be a complete Boolean algebra, and let P be a basis for B, with respect to an order isomorphism e. Identify P with $e[P]$, so that P becomes a dense subset of B. For each B-sentence σ and each $p \in P$ we define the relation *p forces σ*—written $p \Vdash \sigma$—by

$$p \Vdash \sigma \text{ iff } p \leq [\![\sigma]\!]^B.$$

The basic properties of the forcing relation are contained in the final theorem of this section. We write $[\![\sigma]\!]$ for $[\![\sigma]\!]^B$ as usual.

2.5. Theorem. *Let σ and τ be B-sentences and let $\phi(x)$ be a B-formula. Then:*

(i) $p \Vdash \neg \sigma$ iff $\neg \exists q \leq p [q \Vdash \sigma]$;

(ii) $p \Vdash \sigma \wedge \tau$ iff $p \Vdash \sigma$ and $p \Vdash \tau$;

(iii) $p \Vdash \sigma \vee \tau$ iff $\forall q \leq p \exists r \leq q [r \Vdash \sigma$ or $r \Vdash \tau]$;

(iv) $p \Vdash \sigma \to \tau$ iff $\forall q \leq p [q \Vdash \sigma \to \exists r \leq q [r \Vdash \tau]]$;

(v) $p \Vdash \forall x \phi(x)$ iff $\forall u \in V^{(B)} [p \Vdash \phi(u)]$;

(vi) $p \Vdash \exists x \phi(x)$ iff $\forall q \leq p \exists r \leq q \exists u \in V^{(B)} [r \Vdash \phi(u)]$;

(vii) for $a \in V, p \Vdash \forall x \in \hat{a} \phi(x)$ iff $\forall x \in a [p \Vdash \phi(\hat{x})]$;

2. Forcing and Independence Proofs

(viii) *for $a \in V, p \Vdash \exists x \in \hat{a}\phi(x)$ iff $\forall q \leq p \exists r \leq q \exists x \in a[r \Vdash \phi(\hat{x})]$;*

(ix) $[\![\sigma]\!] = 0$ *iff* $\neg \exists p [p \Vdash \sigma]$;

(x) $[\![\sigma]\!] = 1$ *iff* $\forall p [p \Vdash \sigma]$;

(xi) $\forall p \exists q \leq p [q \Vdash \sigma$ or $q \Vdash \neg \sigma]$;

(xii) $[p \Vdash \sigma] \to \neg [p \Vdash \neg \sigma]$;

(xiii) $[q \leq p$ and $p \Vdash \sigma] \to q \Vdash \sigma$.

Proof. We prove some of these assertions, leaving the rest to the reader.

(i) If $p \Vdash \neg \sigma$, then $p \leq [\![\sigma]\!]^*$. So in this case if $q \leq p$, then $q \not\leq [\![\sigma]\!]$ (otherwise $q \leq [\![\sigma]\!]^* \wedge [\![\sigma]\!] = 0$), so $\neg [q \Vdash \sigma]$. Conversely, if $\neg [p \Vdash \neg \sigma]$, then $p \not\leq [\![\sigma]\!]^*$, so $p \wedge [\![\sigma]\!] \neq 0$, so, since P is dense, $\exists q \leq p [q \leq [\![\sigma]\!]]$, whence $\exists q \leq p [q \Vdash \sigma]$.

(ii) $p \Vdash \sigma \wedge \tau$ iff $p \leq [\![\sigma]\!] \wedge [\![\tau]\!]$
 iff $p \leq [\![\sigma]\!]$ and $p \leq [\![\tau]\!]$
 iff $p \Vdash \sigma$ and $p \Vdash \tau$.

(iii) $p \Vdash \sigma \vee \tau$ iff $p \Vdash \neg(\neg \sigma \wedge \neg \tau)$
 iff $\neg \exists q \leq p [q \Vdash \neg \sigma \wedge \neg \tau]$
 iff $\neg \exists q \leq p [q \Vdash \neg \sigma$ and $q \Vdash \neg \tau]$
 iff $\neg \exists q \leq p [\neg \exists r \leq q [r \Vdash \sigma]$ and $\neg \exists r \leq q [r \Vdash \tau]]$
 iff $\forall q \leq p [\exists r \leq q [r \Vdash \sigma]$ or $\exists r \leq q [r \Vdash \tau]]$
 iff $\forall q \leq p \exists r \leq q [r \Vdash \sigma$ or $r \Vdash \tau]$.

(vi) $p \Vdash \exists x \phi(x)$ iff $p \Vdash \neg \forall x \neg \phi(x)$
 iff $\neg \exists q \leq p [q \Vdash \forall x \neg \phi(x)]$
 iff $\neg \exists q \leq p \forall u \in V^{(B)} \neg \exists r \leq q [r \Vdash \phi(u)]$
 iff $\forall q \leq p \exists u \in V^{(B)} \exists r \leq q [r \Vdash \phi(u)]$.

(xi) Either $p \Vdash \sigma$ or $\neg [p \Vdash \sigma]$. If the former, we are finished. If the latter, then $p \not\leq [\![\sigma]\!]$, so $p \wedge [\![\sigma]\!]^* \neq 0$, whence $\exists q [q \leq p \wedge [\![\sigma]\!]^*]$, so $\exists q \leq p [q \Vdash \neg \sigma]$. □

The meaning of Theorem 2.5 can be clarified as follows. We recall that the elements of p are to be thought of as 'states of information', or briefly, 'states'. Also, for $p, q \in P$, the relation $p \leq q$ means that 'state' p is a refinement (of the information in) 'state' q. Then $p \Vdash \sigma$ may be thought of as asserting that in 'state' p we are in definite possession of the 'fact' σ, or have been 'forced' to accept σ as true. Using this approach, the interpretation of, e.g. (i) in Theorem 2.4 is:

$p \Vdash \neg \sigma$ iff *in no state which refines p will we be forced to accept σ*;

that of (vi) is:

$p \Vdash \exists x \phi(x)$ iff *any refinement of p can be itself refined to a state in which we can instantiate $\phi(x)$*;

and that of (xi) is:

each state can be refined to one in which we either accept σ, or we accept $\neg \sigma$.

The reader may provide similar interpretations of the other clauses in Theorem 2.5.

Finally, we point out that the notion of forcing introduced here is what Cohen (1966) called *weak* forcing; Cohen's original notion of forcing (which we shall write $\Vdash_c$) is customarily known as *strong* forcing. The two notions are related by the equivalence

$$p \Vdash \sigma \leftrightarrow p \Vdash_c \neg\neg \sigma. \tag{*}$$

The chief difference between weak and strong forcing is that, while the former obeys all the laws of *classical* logic, the latter obeys only the laws of *intuitionistic* logic. This means, for example, that if $p \Vdash \sigma$ then $p \Vdash \tau$ whenever τ is *classically* equivalent to σ, while if $p \Vdash_c \sigma$, then one can only infer that $p \Vdash_c \tau$ when τ satisfies the stronger condition of being *intuitionistically* equivalent to σ. (Note in this connection the resemblance between (*) and the usual translation of classical into intuitionistic logic.)

Independence of the Axiom of Constructibility and the Continuum Hypothesis

We are now in a position to use the techniques introduced in earlier sections

2. Forcing and Independence Proofs

to prove (among other things) the independence of the axiom of constructibility and the continuum hypothesis from ZFC.

2.6. Theorem. *Let $B = \text{RO}(2^\omega)$. Then:*

(i) $V^{(B)} \models (P\omega)\widehat{} \neq P\hat{\omega}$.

(ii) $V^{(B)} \models P\hat{\omega} \not\subseteq L$.

Proof. Put $P = C(\omega, 2)$; then B is a completion of P, P is a basis for B, and each $p \in P$ is identified with the element

$$N(p) = \{f \in 2^\omega : p \subseteq f\}$$

of B. Define $u \in V^{(B)}$ by $\text{dom}(u) = \text{dom}(\hat{\omega})$ and

$$u(\hat{n}) = \{f \in 2^\omega : f(n) = 1\} \in B.$$

It is easy to verify that, for $p \in P$ and $n \in \omega$, we have

$$p \Vdash \hat{n} \in u \text{ iff } p(n) = 1;$$
$$p \Vdash \hat{n} \notin u \text{ iff } p(n) = 0.$$

Also,

$$[\![u \in P\hat{\omega}]\!]^B = [\![u \subseteq \hat{\omega}]\!]^B = \bigwedge_{n \in \omega} [\![u(\hat{n}) \Rightarrow [\![\hat{n} \in \hat{\omega}]\!]^B]\!] = 1.$$

We claim that $[\![u = \hat{x}]\!]^B = 0$ for all $x \in P\omega$. For suppose not; then there is $p \in P$ and $x \in P\omega$ such that $p \Vdash u = \hat{x}$. Pick $n \in \omega$ such that $n \notin \text{dom}(p)$. (Possible, since $\text{dom}(p)$ is finite!) If $n \in x$, put $p' = p \cup \{\langle n, 0 \rangle\}$; if $n \notin x$, put $p' = p \cup \{\langle n, 1 \rangle\}$. Then, if $n \in x$, we have $p' \Vdash \hat{n} \in \hat{x} \wedge \hat{n} \notin u$, and if $n \notin x$, we have $p' \Vdash \hat{n} \notin \hat{x} \wedge \hat{n} \in u$. Thus in either case $p' \Vdash u \neq \hat{x}$. But since $p' \leq p$, we have $p' \Vdash u = \hat{x}$, which is a contradiction. This establishes the claim.

It follows that

$$[\![u \in (P\omega)\widehat{}]\!]^B = \bigvee_{x \in P\omega} [\![u = \hat{x}]\!]^B = 0,$$

so

$$1 = [\![u \in P\hat{\omega}]\!]^B \wedge [\![u \notin (P\omega)\hat{\,}]\!]^B \leq [\![P\hat{\omega} \neq (P\omega)\hat{\,}]\!]^B$$

and (i) is proved.

(ii) Consider the set $u \in V^{(B)}$ defined in the proof of (i). We have, by 1.46,

$$[\![L(u)]\!]^B = \bigvee_{x \in L} [\![u = \hat{x}]\!]^B$$
$$= \bigvee_{x \in L \cap P\omega} [\![u = \hat{x}]\!]^B \vee \bigvee_{x \in L - P\omega} [\![u = \hat{x}]\!]^B.$$

Since we already know that $[\![u \in P\hat{\omega}]\!]^B = 1$, we have, for $x \notin P\omega$,

$$[\![u = \hat{x}]\!]^B = [\![u = \hat{x}]\!]^B \wedge [\![u \in P\hat{\omega}]\!]^B \leq [\![\hat{x} \in P\hat{\omega}]\!]^B = [\![\hat{x} \subseteq \hat{\omega}]\!]^B = 0$$

since $x \not\subseteq \omega$. Therefore

$$[\![L(u)]\!]^B = \bigvee_{x \in L \cap P\omega} [\![u = \hat{x}]\!]^B \leq \bigvee_{x \in P\omega} [\![u = \hat{x}]\!]^B = [\![u \in (P\omega)\hat{\,}]\!]^B = 0.$$

Hence $[\![L(u)]\!]^B = 0$ and (ii) follows immediately. □

Recall that in the course of proving Theorem 2.6(i) we remarked that, for $p \in P = C(\omega, 2)$, we have $p \Vdash \hat{n} \in u$ iff $p(n) = 1$, $p \Vdash \hat{n} \notin u$ iff $p(n) = 0$. Accordingly, each condition $p \in P$ may be regarded as encoding a finite 'piece of information' about the members of the 'new subset' u of ω. We may therefore think of $C(\omega, 2)$ as *the set of conditions for adjoining a new subset of ω using finite pieces of information*, and its completion $\text{RO}(2^\omega)$ as *an algebra which adjoins a new subset of ω*.

1.19, 1.33 and 2.6 now give:

2.7. Corollary. *If* ZF *is consistent, so is* ZFC+ *'there is a non-constructible subset of ω'*. □

We next show how to extend this result to include the GCH.

2.8. Theorem. *Assume the* GCH. *Then, if B satisfies* ccc *and* $|B| = 2^{\aleph_0}$,

$$V^{(B)} \models \text{GCH}.$$

2. Forcing and Independence Proofs

Proof. Recall from 1.39 that we have, for any $u \in V^{(B)}$,

$$\mathrm{dom}(P^{(B)}(u)) = B^{\mathrm{dom}(u)}$$

and

$$V^{(B)} \models P^{(B)}(u) = Pu.$$

Take $u = \hat{\aleph}_\alpha$. Then since the map $x \mapsto \hat{x}$ is one-one, we have $|\mathrm{dom}(\hat{\aleph}_\alpha)| = \aleph_\alpha$. Thus, since the GCH is assumed to hold,

$$|\mathrm{dom}(P^{(B)}(\hat{\aleph}_\alpha))| = |B^{\mathrm{dom}(\hat{\aleph}_\alpha)}| = (2^{\aleph_0})^{\aleph_\alpha} = 2^{\aleph_\alpha} = \aleph_{\alpha+1}.$$

It follows from 1.52 that

$$V^{(B)} \models |P^{(B)}(\hat{\aleph}_\alpha)| \leq |\hat{\aleph}_{\alpha+1}|.$$

But B satisfies ccc, so, by 1.51, for any α,

$$V^{(B)} \models \hat{\aleph}_\alpha = \aleph_{\hat{\alpha}}$$

whence

$$V^{(B)} \models |\hat{\aleph}_{\alpha+1}| = \aleph_{\hat{\alpha}+1}.$$

These two facts give

$$V^{(B)} \models |P^{(B)}(\aleph_{\hat{\alpha}})| \leq \aleph_{\hat{\alpha}+1},$$

so that

$$V^{(B)} \models |P\aleph_{\hat{\alpha}}| \leq \aleph_{\hat{\alpha}+1}.$$

Since this holds for *arbitrary* α, it follows using 1.45 that

$$V^{(B)} \models \forall \alpha [|P\aleph_\alpha| \leq \aleph_{\alpha+1}]$$

and so

$$V^{(B)} \models \mathrm{GCH}. \quad \square$$

2.9. Corollary. *If* ZF *is consistent, so is* ZFC + GCH + *'there is a non-constructible subset of ω'.*

Proof. Let B be the algebra introduced in 2.6. Then B satisfies ccc and $|B| = 2^{\aleph_0}$. Since $Consis(\text{ZF}) \to Consis(\text{ZFC} + \text{GCH})$, the required result now follows immediately from 2.6, 2.8, 1.33 and 1.19. □

We now turn to the problem of violating the continuum hypothesis in $V^{(B)}$. The idea here is to make $P^{(B)}(\hat{\omega})$ large in $V^{(B)}$; we shall see that this can be achieved by taking an appropriate B of large cardinality. On the other hand, if we want to pin down the cardinality of $P^{(B)}(\hat{\omega})$ in $V^{(B)}$, we shall need to make a reasonably precise estimate of $|B|$. We now set about doing this for the sort of B we have in mind.

A topological space X is said to satisfy the *countable chain condition* (ccc) if each disjoint family of sets open in X is countable. We have already remarked that the product space 2^I satisfies ccc.

2.10. Lemma. *Let X be a topological space satisfying* ccc. *Let E be a base for X and let B be the regular open algebra of X. Then $|B| \leq |E|^{\aleph_0}$.*

Proof. Let $U \in B$, and, using Zorn's lemma, let F be a maximal disjoint subfamily of $E \cap PU$. Put $G = \bigcup F$. We claim that $U = (\overline{G})^\circ$. For since $G \subseteq U$ and U is regular open, we have $(\overline{G})^\circ \subseteq (\overline{U})^\circ = U$. On the other hand, consider $U - \overline{G}$. This is an open set; if it is non-empty then it includes a non-empty member of E which is disjoint from every member of F, contradicting the maximality of F. Thus $U - \overline{G} = \emptyset$, so that $U \subseteq \overline{G}$ and $U \subseteq (\overline{G})^\circ$. This proves the claim.

Accordingly each member of B is determined by a disjoint subfamily of E; since X satisfies ccc each such subfamily is countable and there are at most $|E|^{\aleph_0}$ of them. □

2.11. Corollary. *For each set I let 2^I be the product space where 2 is assigned the discrete topology. If $|I| = \aleph_\alpha$, then*

$$\aleph_\alpha \leq |\text{RO}(2^I)| \leq \aleph_\alpha^{\aleph_0}.$$

2. Forcing and Independence Proofs

Proof. The family of sets of the form

$$\{f \in 2^I : f(i_1) = a_1, \ldots, f(i_n) = a_n\}$$

where $i_1, \ldots, i_n \in I$ and $a_1, \ldots, a_n \in 2$ is a base for 2^I of cardinality $\aleph_\alpha$. Since each set of this form is clopen, it is in $\mathrm{RO}(2^I)$ and so $\aleph_\alpha \leq |\mathrm{RO}(2^I)|$. On the other hand, 2^I satisfies ccc and so 2.10 applies to yield the other inequality. □

We are now in a position to prove

2.12. Theorem. *Suppose that $\aleph_\alpha^{\aleph_0} = \aleph_\alpha$ and let $B = \mathrm{RO}(2^{\omega \times \omega_\alpha})$. Then*

$$V^{(B)} \models 2^{\aleph_0} = \aleph_{\hat\alpha}.$$

Proof. By 2.11 we have

$$\aleph_\alpha \leq |B| \leq \aleph_\alpha^{\aleph_0} = \aleph_\alpha,$$

so that $|B| = \aleph_\alpha$. Hence

$$|\mathrm{dom}(P^{(B)}(\hat\omega))| = |B^{\mathrm{dom}(\hat\omega)}| = \aleph_\alpha^{\aleph_0} = \aleph_\alpha$$

and so, by 1.52

$$V^{(B)} \models |P^{(B)}(\hat\omega)| \leq |\hat{\aleph}_\alpha|,$$

whence

$$V^{(B)} \models |P\omega| \leq |\hat{\aleph}_\alpha|.$$

But B satisfies ccc, so by 1.51 we have $V^{(B)} \models |\hat{\aleph}_\alpha| = \aleph_{\hat\alpha}$, whence

$$V^{(B)} \models |P\omega| \leq \aleph_{\hat\alpha},$$

i.e.

$$V^{(B)} \models 2^{\aleph_0} \leq \aleph_{\hat\alpha}.$$

It remains to show that

$$V^{(B)} \models \aleph_{\hat{\alpha}} \leq 2^{\aleph_0}.$$

To this end, for each $\nu < \omega_\alpha$ define $u_\nu \in V^{(B)}$ by $\text{dom}(u_\nu) = \text{dom}(\hat{\omega})$ and

$$u_\nu(\hat{n}) = \{f \in 2^{\omega \times \omega_\alpha} : f(n,\nu) = 1\}.$$

We have

$$[\![u_\nu \subseteq \hat{\omega}]\!]^B = \bigwedge_{n \in \omega} [u_\nu(\hat{n}) \Rightarrow [\![\hat{n} \in \hat{\omega}]\!]^B] =: 1.$$

We also know that $P = C(\omega \times \omega_\alpha, 2)$ is a basis for B. Moreover, it is easy to verify that, for $p \in P$,

$$p \Vdash \hat{n} \in u_\nu \text{ iff } p(n,\nu) = 1;$$
$$p \Vdash \hat{n} \notin u_\nu \text{ iff } p(n,\nu) = 0.$$

We claim that, if $\mu, \nu < \omega_\alpha$ and $\mu \neq \nu$, then $[\![u_\mu = u_\nu]\!]^B = 0$. For suppose not; then there are $\mu, \nu < \omega_\alpha$, $\mu \neq \nu$ and $p \in P$ such that $p \Vdash u_\mu = u_\nu$. Choose $n \in \omega$ so that $\langle n, \xi \rangle \notin \text{dom}(p)$ for any $\xi < \omega_\alpha$ (possible, since $\text{dom}(p)$ is finite!) and put

$$p' = p \cup \{\langle\langle n, \mu\rangle, 1\rangle\} \cup \{\langle\langle n, \nu\rangle, 0\rangle\}.$$

Then $p' \Vdash \hat{n} \in u_\mu \wedge \hat{n} \notin u_\nu$, whence $p' \Vdash u_\mu \neq u_\nu$. But since $p' \leq p$ and $p \Vdash u_\mu = u_\nu$, it follows that $p' \Vdash u_\mu = u_\nu$. This contradiction proves the claim.

Now define $f \in V^{(B)}$ by

$$f = \{\langle \hat{\nu}, u_\nu \rangle^{(B)} : \nu < \omega_\alpha\} \times \{1\}.$$

One then easily verifies (cf. proof of 1.52) that

$$V^{(B)} \models f \text{ is a map of } \hat{\omega}_\alpha \text{ into } P\hat{\omega}.$$

Moreover, since $[\![u_\mu = u_\nu]\!]^B = 0$ for $\mu \neq \nu$, it quickly follows that $V^{(B)} \models f$ is one-one. Since B satisfies the ccc, we have, by 1.51, $V^{(B)} \models \hat{\omega}_\alpha = \omega_{\hat{\alpha}}$,

2. Forcing and Independence Proofs

so that

$$V^{(B)} \models f \text{ is a one-one map of } \omega_{\hat{\alpha}} \text{ into } P\hat{\omega}.$$

Hence $V^{(B)} \models \aleph_{\hat{\alpha}} \leq 2^{\aleph_0}$, and we are done. □

In the proof of this last theorem we remarked that, for $p \in P = C(\omega \times \omega_\alpha, 2)$, we have $p \Vdash \hat{n} \in u_\nu$ iff $p(n,\nu) = 1$; $p \Vdash \hat{n} \in u_\nu$ iff $p(n,\nu) = 0$. Thus each condition $p \in P$ may be thought of as encoding a finite 'piece of information' about the members of the $\aleph_\alpha$ 'new subsets' u_ν of ω. We may therefore regard $C(\omega \times \omega_\alpha, 2)$ as *the set of conditions for adjoining $\aleph_\alpha$ new subsets of ω using finite pieces of information,* and its completion $RO(2^{\omega \times \omega_\alpha})$ as an *algebra which adjoins $\aleph_\alpha$ new subsets of ω.*

2.13. Corollary. *If* ZF *is consistent, so is* ZFC $+ 2^{\aleph_0} = \aleph_2$.

Proof. In ZFC + GCH we have $\aleph_2^{\aleph_0} = (2^{\aleph_1})^{\aleph_0} = 2^{\aleph_1} = \aleph_2$, so by 2.12 one can prove in ZFC + GCH the existence of a complete Boolean algebra B such that $V^{(B)} \models 2^{\aleph_0} = \aleph_2$. Since $Consis(\text{ZF}) \to Consis(\text{ZFC} + \text{GCH})$, the required result now follows from 1.33 and 1.19. □

Similar arguments show that in 2.13 '$\aleph_2$' can be replaced by '$\aleph_3$', '$\aleph_{\omega+1}$', '$\aleph_{\omega_1}$', etc.

Problems

Throughout, κ denotes an infinite cardinal and B a complete Boolean algebra. In Problem 2.14, Problem 2.15 and Problem 2.16, for any cardinal λ, λ^κ and $\hat{\lambda}^{\hat{\kappa}}$ are understood to denote, respectively, the set of all maps of κ into λ, and 'the set of all maps of $\hat{\kappa}$ into $\hat{\lambda}$ in $V^{(B)}$'.

2.14. (*Infinite distributive laws and $V^{(B)}$*). Let λ be a cardinal (finite or infinite). B is said to be (κ, λ)-*distributive* if for any double sequence $\{b_{\alpha\beta} : \langle\alpha,\beta\rangle \in \kappa \times \lambda\} \subseteq B$ we have

$$\bigwedge_{\alpha<\kappa} \bigvee_{\beta<\lambda} b_{\alpha\beta} = \bigvee_{f \in \lambda^\kappa} \bigwedge_{\alpha<\kappa} b_{\alpha f(\alpha)}.$$

Show that the following conditions are equivalent:

(i) B is (κ, λ)-distributive;

(ii) $V^{(B)} \models \hat{\lambda}^{\hat{\kappa}} = (\lambda^\kappa)\hat{\ }$.

(Use the fact that
$$[\![h \in \hat{\lambda}^{\hat{\kappa}}]\!] = \bigwedge_{\alpha<\kappa} \bigvee_{\beta<\lambda} b_{\alpha\beta} \text{ and } [\![h \in (\lambda^\kappa)\hat{\ }]\!] = \bigvee_{f\in\lambda^\kappa} \bigwedge_{\alpha<\kappa} b_{\alpha f(\alpha)},$$
where $b_{\alpha\beta} = [\![h(\hat{\alpha}) = \hat{\beta}]\!]$.)

2.15. (*Infinite distributive laws and $V^{(B)}$ continued*).

(i) B is said to satisfy the *restricted $(\kappa, 2)$-distributive law* if for each double sequence $\{b_{\alpha n} : \langle \alpha, n \rangle \in \kappa \times 2\} \subseteq B$ such that $b_{\alpha 0} = b_{\alpha 1}^*$ for all $\alpha < \kappa$, we have $\bigvee_{f \in 2^\kappa} \bigwedge_{\alpha<\kappa} b_{\alpha f(\alpha)} = 1$.

Show that the following conditions are equivalent:

(a) B satisfies the restricted $(\kappa, 2)$-distributive law;

(b) B is $(\kappa, 2)$-distributive;

(c) B is (κ, κ)-distributive;

(d) B is $(\kappa, 2^\kappa)$-distributive;

(e) $V^{(B)} \models \hat{\kappa}^{\hat{\kappa}} = (\kappa^\kappa)\hat{\ }$;

(f) $V^{(B)} \models P\hat{\kappa} = (P\kappa)\hat{\ }$.

(Show that (a) → (f) → (e) → (d) → (c) → (b) → (a). For (a) → (f), use (a) to prove $V^{(B)} \models u \in (P\kappa)\hat{\ }$ for any $u \in B^{\text{dom}(\hat{\kappa})}$. For (f) → (e), use the fact that $\kappa^\kappa \subseteq 2^{\kappa \times \kappa}$ and $|\kappa| = |\kappa \times \kappa|$. For (e) → (d), observe that $(2^\kappa)^\kappa = 2^{\kappa \times \kappa}$ and use 2.12. The remaining implications are trivial.)

(ii) Show that $RO(2^\omega)$ is not $(\omega, 2)$-distributive. (Use (i) and 2.6(i).)

2.16. (*Weak distributive laws and $V^{(B)}$*). B is said to be *weakly (ω, κ)-distributive* if for each double sequence $\{b_{n\alpha} : \langle n, \alpha \rangle \in \omega \times \kappa\}$ we have
$$\bigwedge_{n\in\omega} \bigvee_{\alpha<\kappa} b_{n\alpha} = \bigvee_{f\in\kappa^\omega} \bigwedge_{n\in\omega} \bigvee_{\alpha \leq f(n)} b_{n\alpha}.$$

2. Forcing and Independence Proofs

Clearly, if B is (ω, κ)-distributive, B is weakly (ω, κ)-distributive. (But the converse fails: for example, if B is the complete Boolean algebra of Lebesgue measurable subsets of $[0, 1]$ modulo the ideal of sets of measure 0, then it can be shown that B is weakly (ω, ω)-distributive but not (ω, ω)-distributive.)

We define the *cofinality* $\text{cf}(\kappa)$ of κ to be the least ordinal which is the order type of a cofinal subset of κ. Thus $\text{cf}(\kappa) > \omega$ iff $\forall f \in \kappa^{\omega} \exists \beta < \kappa \forall n \in \omega [f(n) \leq \beta]$, i.e. iff each function from ω to κ is bounded by some ordinal $< \kappa$.

(i) Show that, if $\text{cf}(\kappa) > \omega$, then

$$\bigvee_{f \in \kappa^{\omega}} \bigwedge_{n \in \omega} \bigvee_{\alpha \leq f(n)} b_{n\alpha} = \bigvee_{\beta < \kappa} \bigwedge_{n \in \omega} \bigvee_{\alpha \leq \beta} b_{n\alpha}.$$

(ii) Show that, if B satisfies ccc and $\text{cf}(\kappa) > \omega$, then B is weakly (ω, κ)-distributive. (Let $\{b_{n\alpha} : \langle n, \alpha \rangle \in \omega \times \kappa\} \subseteq B$. Using 1.53(iv), replace each $\{b_{n\alpha} : \alpha < \kappa\}$ by a countable subset C_n having the same supremum.)

(iii) Suppose that $\text{cf}(\kappa) > \omega$. Show that B is weakly (ω, κ)-distributive iff $V^{(B)} \models (\text{cf } \kappa)\hat{\ } > \hat{\omega}$. (Note that $[\![f \in \hat{\kappa}^{\hat{\omega}}]\!] = \bigwedge_{n \in \omega} \bigvee_{\alpha < \kappa} b_{n\alpha}$ and that $[\![\exists \beta < \hat{\kappa} \forall n \in \hat{\omega}[f(n) \leq \beta]]\!] = \bigvee_{\beta < \kappa} \bigwedge_{n \in \omega} \bigvee_{\alpha \leq \beta} b_{n\alpha}$, where $b_{n\alpha} = [\![f(\hat{n}) = \hat{\alpha}]\!]$.)

(iv) Show that B is weakly (ω, ω)-distributive iff $V^{(B)} \models \forall g[g \in \hat{\omega}^{\hat{\omega}} \to \exists f \in (\omega^{\omega})\hat{\ } \forall n \in \hat{\omega}[g(n) \leq f(n)]]$, in other words, iff in $V^{(B)}$ the *standard* numerical functions are cofinal in the class of *all* numerical functions. (Argue as in (iii).)

2.17. (κ-*closure and* $V^{(B)}$). Let P be a basis for B. P is said to be κ-*closed* if for each ordinal $\alpha < \kappa$ and each descending α-sequence $p_0 \geq p_1 \geq \ldots \geq p_\beta \geq \ldots (\beta < \alpha)$ in P there is $p \in P$ such that $p \leq p_\beta$ for all $\beta < \alpha$. Consider the following conditions:

(i) B has a dense subset P which is κ-closed;

(ii) for any $\alpha < \kappa$ and any $x \in V$,

$$V^{(B)} \models \hat{x}^{\hat{\alpha}} = (x^{\alpha})\hat{\ };$$

(iii) $V^{(B)} \models \mathrm{Card}(\hat{\alpha})$ for any cardinal $\alpha \leq \kappa$;

(iv) $V^{(B)} \models P\hat{\alpha} = (P\alpha)\check{}$ for any $\alpha < \kappa$.

Show that (i) $\to$ (ii) $\to$ (iii), and (ii) $\to$ (iv). Hence (i) implies that B is (α, λ)-distributive for any $\alpha < \kappa$ and any λ. (For (i) $\to$ (ii), let $p \in P$ be such that $p \Vdash f \in \hat{x}^{\hat{\alpha}}$. Using (i), find a descending sequence $\{p_\beta : \beta < \alpha\} \subseteq P$ and a set $\{y_\beta : \beta < \alpha\} \subseteq x$ such that $p_\beta \Vdash f(\hat{\beta}) = \hat{y}_\beta$. If $q \in P$ satisfies $q \leq p_\beta$ for all $\beta < \alpha$, and $g = \{\langle \beta, y_\beta \rangle : \beta < \alpha\}$, show that $q \Vdash f = \hat{g}$.)

2.18. (*An important set of conditions*). Let x and y be non-empty sets, where $|y| \geq 2$. We put $C_\kappa(x, y)$ for the set—partially ordered by $\supseteq$—of all maps with domain a subset of x of cardinality $< \kappa$ and range a subset of y. Put $B_\kappa(x, y)$ for the regular open algebra of the space y^x with the topology whose basic open sets are of the form

$$N(p) = \{f \in y^x : p \subseteq f\}$$

for $p \in C_\kappa(x, y)$. Observe that $C_\omega(x, y) = C(x, y)$.

(i) Show that $C_\kappa(x, y)$ is refined, and that $\langle B_\kappa(x, y), N \rangle$ is a Boolean completion of $C_\kappa(x, y)$. Thus $C_\kappa(x, y)$ is a basis for $B_\kappa(x, y)$.

(ii) Show that, if κ is regular, $C_\kappa(x, y)$ is κ-closed.

(iii) Assume GCH, κ is regular and $|y| \leq \kappa$. Show that, if I is any set of pairwise incompatible elements of $C_\kappa(x, y)$, then $|I| \leq \kappa$. (Fix a well-ordering of I. Define a sequence $\{x_\alpha : \alpha < \kappa^+\}$ of subsets of x by: $x_0 = \emptyset$; $x_\alpha = \bigcup_{\beta < \alpha} x_\beta$ for limit α, $x_{\alpha+1} = x_\alpha \cup \bigcup \{\mathrm{dom}(q) :$ for some $p \in C_\kappa(x_\alpha, y)$, q is the least element of I such that $q|x_\alpha = p\}$. Show that $|x_\alpha| \leq \kappa$ for $\alpha < \kappa^+$, and hence that $|C_\kappa(x_\kappa, y)| \leq \kappa$. Now prove that $I \subseteq C_\kappa(x_\kappa, y)$: for any $p \in I$, show that there is $\alpha < \kappa$ such that $\mathrm{dom}(p) \cap x_\alpha = \mathrm{dom}(p) \cap x_{\alpha+1}$; choose $q \in I$ such that $p|x_\alpha = q|x_\alpha$ and $\mathrm{dom}(q) \subseteq x_{\alpha+1}$; show that p and q coincide on $\mathrm{dom}(p) \cap \mathrm{dom}(q)$; deduce that $p = q$ and $p \in C_\kappa(x_\kappa, y)$.)

(iv) Assume GCH, κ is regular and $|y| \leq \kappa$. Show that $B_\kappa(x, y)$ satisfies $\kappa^+ - \mathrm{cc}$. (Use (iii).)

2. Forcing and Independence Proofs

(v) Assume GCH, κ and $|x|$ are regular and $\kappa < |x|$. Show that $|B_\kappa(x,2)| = |x|$. (Using (iii), argue as in 2.10.)

2.19. (*Consistency of* $2^{\aleph_0} = \aleph_2 + \forall \kappa \geq \aleph_1 [2^\kappa = \kappa^+]$ *with* ZFC).

(i) Show that, if $|B| = \lambda$, then $V^{(B)} \models |P\hat{\kappa}| \leq |(\lambda^\kappa)\hat{}\,|$. (Use 1.52.)

*(ii)† Assume GCH, and let $|B| = \lambda \geq \aleph_0$. Show that $V^{(B)} \models \forall \alpha \geq \hat{\lambda}[\text{Card}(\alpha) \to 2^\alpha = \alpha^+]$. (Use 1.53 and (i).)

(iii) Assume GCH, and let $B = \text{RO}(2^{\omega \times \omega_2})$. Show that $V^{(B)} \models 2^{\aleph_1} = \aleph_2$ (use (i)) and deduce from this, 2.12 and (ii) that if ZF is consistent, so is ZFC $+ 2^{\aleph_0} = \aleph_2 + \forall \kappa \geq \aleph_1[2^\kappa = \kappa^+]$.

2.20. (*A further relative consistency result*). Assume GCH. Let κ, λ be regular cardinals such that $\kappa < \lambda$. Put $B = B_\kappa(\kappa \times \lambda, 2)$ (2.18).

(i) Show that $V^{(B)} \models \text{Card}(\hat{\alpha})$ for any cardinal α. (For $\alpha \leq \kappa$, use 2.18 (ii) and 2.17(iii). For $\alpha \geq \kappa^+$, use 2.18(iv) and 1.53).

(ii) Show that $V^{(B)} \models P\hat{\alpha} = (P\alpha)\hat{}\,$ for any cardinal $\alpha < \kappa$. (Use 2.18(ii) and 2.17(iv).)

(iii) Show that
$$V^{(B)} \models \forall \alpha < \hat{\kappa}[\text{Card}(\alpha) \wedge \aleph_0 \leq \alpha \to 2^\alpha = \alpha^+].$$
(Use (i), (ii) and GCH.)

(iv) Show that
$$V^{(B)} \models \forall \alpha \geq \hat{\lambda}[\text{Card}(\alpha) \to 2^\alpha = \alpha^+].$$
(Use 2.18(v), 2.19(i) and (i).)

(v) Show that
$$V^{(B)} \models \forall \alpha[\text{Card}(\alpha) \wedge \hat{\kappa} \leq \alpha < \hat{\lambda} \to 2^\alpha = \hat{\lambda}].$$

---

†This result shows that, as long as B is a *set* (*i.e.* has a cardinality) we cannot *provably* violate the GCH in $V^{(B)}$ at arbitrarily high cardinals. It turns out, however, that this can be achieved when B is a suitably chosen (proper) *class*. The details are, unfortunately, too lengthy to be included here. See Easton (1970), Takeuti and Zaring (1973) or Shoenfield (1971).

(Use 2.18(iv) and 2.19(i) to show that $V^{(B)} \models 2^{\hat{\alpha}} \leq \hat{\lambda}$ for $\kappa \leq \alpha < \lambda$ and argue as in the proof of 2.12 to get $V^{(B)} \models 2^{\hat{\kappa}} \geq \hat{\lambda}$.)

(vi) Deduce that, if ZF is consistent, so is

$$\text{ZFC} + 2^{\aleph_0} = \aleph_1 + \forall \kappa [\aleph_1 \leq \kappa \leq \aleph_\omega \to 2^\kappa = \aleph_{\omega+1}]$$
$$+ \forall \kappa [\aleph_{\omega+1} \leq \kappa \to 2^\kappa = \kappa^+].$$

2.21. (*Consistency of* $\text{GCH} + P\omega \subseteq L + P\omega_1 \not\subseteq L$ *with* ZFC). Assume GCH, let κ be regular, and put $B = B_\kappa(\kappa, 2)$.

(i) Show that $V^{(B)} \models \text{Card}(\hat{\alpha})$ for any cardinal α. (Like 2.20(i).)

(ii) Show that $V^{(B)} \models P\hat{\alpha} = (P\alpha)\hat{\,}$ for any cardinal $\alpha < \kappa$. (Like 2.20(ii).)

(iii) Show that $V^{(B)} \models P\hat{\kappa} \neq (P\kappa)\hat{\,}$. (Like 2.6(i).)

(iv) Show that $V^{(B)} \models \text{GCH}$. (To show $V^{(B)} \models 2^{\hat{\alpha}} = \hat{\alpha}^+$ for $\alpha < \kappa$, argue as in 2.20(iii). (To show $V^{(B)} \models 2^{\hat{\alpha}} = \hat{\alpha}^+$ for $\alpha \geq \kappa$, argue as in 2.20(iv).)

(v) Assume $V = L$. Show that

$$V^{(B)} \models \forall \alpha < \hat{\kappa}[P\alpha \subseteq L] \wedge P\hat{\kappa} \not\subseteq L.$$

(Using (i) and (ii), argue as in 2.6.)

(vi) Deduce that, if ZF is consistent, so is $\text{ZFC} + \text{GCH} + P\omega \subseteq L + P\omega_1 \not\subseteq L$.

Chapter 3

Group Actions on $V^{(B)}$ and the Independence of the Axiom of Choice

Group Actions on $V^{(B)}$

Let G be a group, and X a class. An *action* of G on X is a map $\langle g, x \rangle \mapsto g.x \colon G \times X \to X$ satisfying

$$1.x = x \quad \text{and} \quad (gh).x = g.(h.x)$$

for all $x \in X$, $g, h \in G$, where 1 is the identity element of G. (When confusion is unlikely, we write gx for $g.x$.) Under these conditions we say that *G acts* on X. For each $g \in G$, the map $\pi_g \colon X \to X$ defined by $\pi_g(x) = g.x$ is a permutation of X, and the correspondence $g \mapsto \pi_g$ defines a homomorphism of G into the group of permutations of X.

If B is a Boolean algebra, by an action of a group G on B we shall always mean an *action of G by automorphisms*, that is, one in which each π_g as defined above is not merely a permutation but actually an *automorphism* of B. In particular, the automorphism group $\mathrm{Aut}(B)$ of B acts on B in the natural way via:

$$\pi.b = \pi(b)$$

for $\pi \in \mathrm{Aut}(B), b \in B$.

We can extend the notion of group actions to *Boolean-valued structures* as follows. Let B be a complete Boolean algebra, and let $S =$

$\langle S, [\![.E.]\!]_S, [\![. \in .]\!]_S \rangle$ be a B-valued structure. An *action* of a group G on S is a pair of actions of G on B and the class S satisfying

$$[\![gu = gv]\!]_S = g.[\![u = v]\!]_S \qquad (3.1)$$
$$[\![gu \in gv]\!]_S = g.[\![u \in v]\!]_S .$$

It is easily shown by induction on the complexity of formulas that for any formula $\phi(v_1, \ldots, v_n)$ of L, any $x_1, \ldots, x_n \in S$ and any $g \in G$,

$$[\![\phi(gx_1, \ldots, gx_n)]\!]_S = g.[\![\phi(x_1, \ldots, x_n)]\!]_S . \qquad (3.2)$$

We now show that any action of a group G on a complete Boolean algebra B extends naturally to an action of G on the B-valued structure $V^{(B)}$.

3.3. Theorem. *Let G be a group acting on the complete Boolean algebra B. Define the map $(g, u) \mapsto gu : G \times V^{(B)} \to V^{(B)}$ by recursion on the well-founded relation $y \in \mathrm{dom}(x)$ via*

$$gx = \{\langle gx, g.u(x)\rangle : x \in \mathrm{dom}(u)\} . \qquad (3.4)$$

Then this defines an action of G on $V^{(B)}$ such that:

(i) *for any $u \in V^{(B)}$, $g \in G$, we have $\mathrm{dom}(gu) = \{gx : x \in \mathrm{dom}(u)\}$ and, for any $x \in \mathrm{dom}(u)$, $(gu)(gx) = g.u(x)$;*

(ii) *$g\hat{v} = \hat{v}$ for any $v \in V$.*

Proof. To show that (3.4) defines an action of G on $V^{(B)}$, we first prove that for any $g \in G$ the map $x \mapsto gx$ for $x \in V^{(B)}$ is one-one and sends $V^{(B)}$ into itself. For this it suffices to show by induction on α that

(1) *the restriction to $V_\alpha^{(B)}$ of $x \mapsto gx$ is one-one from $V_\alpha^{(B)}$ to $V^{(B)}$.*

Assume that (1) holds for all $\beta < \alpha$. If $u \in V_\alpha^{(B)}$ then $\mathrm{dom}(u) \subseteq V_\beta^{(B)}$ for some $\beta < \alpha$, so that the restriction of $x \mapsto gx$ to $\mathrm{dom}(u)$ is a one-one map of $\mathrm{dom}(u)$ into $V^{(B)}$. It follows immediately that gu as defined by

3. Group Actions and Independence of AC

(3.4) is a map of $\{gx : x \in \mathrm{dom}(u)\} \subseteq V^{(B)}$ into B, so that $gu \in V^{(B)}$. Thus the restriction to $V_\alpha^{(B)}$ of $x \mapsto gx$ carries $V_\alpha^{(B)}$ into $V^{(B)}$. To show that it is one-one, suppose that $u, v \in V_\alpha^{(B)}$ and $gu = gv$. Then, by (3.4),

(2) $\quad \{\langle gx, g.u(x)\rangle : x \in \mathrm{dom}(u)\} = \{\langle gy, g.v(y)\rangle : y \in \mathrm{dom}(v)\}.$

But there is $\beta < \alpha$ such that $V_\beta^{(B)}$ includes both $\mathrm{dom}(u)$ and $\mathrm{dom}(v)$, so that $x \mapsto gx$ is one-one on both these sets. It now follows from (2) that

$$\{\langle x, u(x)\rangle : x \in \mathrm{dom}(u)\} = \{\langle y, v(y)\rangle : y \in \mathrm{dom}(v)\}$$

i.e. $u = v$. Hence the restriction to $V_\alpha^{(B)}$ of $x \mapsto g.x$ is one-one and (1) is proved.

Part (i) now follows immediately from (3.4).

To establish that $\langle g, u\rangle \mapsto gu$ is an action of G on $V^{(B)}$, we first use the induction principle for $V^{(B)}$ to show that $(gh)u = g(hu)$ for any $g, h \in G$, $u \in V^{(B)}$. Assuming accordingly that $(gh)x = g(hx)$ for all $x \in \mathrm{dom}(u)$, we compute

$$\begin{aligned} g(hu) &= \{\langle gy, g(hu)(y)\rangle : y \in \mathrm{dom}(hu)\} \\ &= \{\langle g(hx), g(hu)(hx)\rangle : x \in \mathrm{dom}(u)\} \\ &= \{\langle g(hx), g(h.u(x))\rangle : x \in \mathrm{dom}(u)\} \\ &= \{\langle (gh)x, (gh).u(x)\rangle : x \in \mathrm{dom}(u)\} \\ &= (gh)u \end{aligned}$$

which proves the assertion. Similarly, one shows that $1u = u$ for all $u \in V^{(B)}$. The facts that

$$g.[\![u \in v]\!] = [\![gu \in gv]\!]$$

and

$$g.[\![u = v]\!] = [\![gu = gv]\!]$$

are proved simultaneously using the induction principle for $V^{(B)}$. We omit the straightforward details.

Finally (ii) is proved by a simple induction on the well-founded relation $y \in x$. $\square$

Recall that the automorphism group $\text{Aut}(B)$ of B acts on B; hence it also acts on $V^{(B)}$. An element x of B or of $V^{(B)}$ is said to be *invariant* if $\pi x = x$ for every $\pi \in \text{Aut}(B)$. B is said to be *homogeneous* if 0 and 1 are its only invariant elements.

3.5. Problem. (*Another characterization of homogeneity*). Show that B is homogeneous iff for each $x \neq 0$, $y \neq 0$ in B there is an automorphism π of B such that $x \wedge \pi y \neq 0$. (Consider $\bigvee \{\pi y : \pi \text{ an automorphism of } B\}$.)

Next we show that homogeneity of B confers certain desirable properties on $V^{(B)}$.

3.6. Lemma. *Suppose that B is homogeneous. Then, for any formula $\phi(v_1, \ldots, v_n)$ and any $x_1, \ldots, x_n \in V$, either $[\![\phi(\hat{x}_1, \ldots, \hat{x}_n)]\!]^B = 0$ or else $[\![\phi(\hat{x}_1, \ldots, \hat{x}_n)]\!]^B = 1$. In particular, for any sentence σ, either $[\![\sigma]\!]^B = 0$ or $[\![\sigma]\!]^B = 1$.*

Proof. By 3.2 and 3.3(ii) $[\![\phi(\hat{x}_1, \ldots, \hat{x}_n)]\!]^B$ is an invariant element of B. The result now follows from the homogeneity of B. □

To conclude this section, we establish the existence of a large class of homogeneous algebras.

3.7. Lemma. *For any set I, $\text{RO}(2^I)$ is homogeneous.*

Proof. For each $i \in I$ define $\pi_i : 2^I \to 2^I$ by $\pi_i f = f^i$ for $f \in 2^I$, where $f^i \in 2^I$ is defined by
$$f^i(j) = f(j) \text{ for } j \neq i$$
$$f^i(i) = 1 - f(i).$$
It is easy to check that π_i is a homeomorphism of 2^I onto itself, and so induces an automorphism π'_i of $\text{RO}(2^I) = B$ defined by
$$\pi'_i(U) = \pi_i^{-1}[U]$$
for $U \in B$.

Suppose now that U is an invariant element of B. Then if $U \neq \emptyset$ there is a basic open set
$$\bigcap_{k=1}^{n} \{f \in 2^I : f(i_k) = a_k\} \subseteq U, \tag{1}$$

3. Group Actions and Independence of AC

where $\{i_1,\ldots,i_n\} \subseteq I$ and $\{a_1,\ldots,a_n\} \subseteq 2$. Applying π'_{i_n} and using the invariance of U, it follows that

$$\bigcap_{k=1}^{n-1} \{f \in 2^I : f(i_k) = a_k\} \cap \{f \in 2^I : f(i_n) = 1 - a_n\} \subseteq U.$$

This, together with (1) implies

$$\bigcap_{k=1}^{n-1} \{f \in 2^I : f(i_k) = a_k\} \subseteq U.$$

Continuing in this way we get $2^I \subseteq U$, so that $U = 1$ in B. The homogeneity of B follows. □

The Independence of the Existence of Definable Well-Orderings of $P\omega$

We now apply the results of the previous section to show that, if ZF is consistent, so is ZFC + GCH + 'there is no definable well-ordering of $P\omega$'. Thus, although in ZFC one can prove the *existence* of a well-ordering of $P\omega$, even in the presence of GCH it is consistent to assume that no such well-ordering can be *explicitly defined*.

For each formula $\phi(x,y)$ let WO_ϕ be the sentence 'ϕ defines a well-ordering of $P\omega$'. Then we have

3.8. Theorem. *Let $B = \text{RO}(2^\omega)$. Then for any formula $\phi(x,y)$ we have*

$$V^{(B)} \models \neg \text{WO}_\phi.$$

Proof. Before launching into formalities we give an outline of the proof. By 2.6 $P\hat\omega - (P\omega)\hat{\,}$ is non-empty in $V^{(B)}$. If $P\hat\omega$ had a definable well-ordering in $V^{(B)}$, $P\hat\omega - (P\omega)\hat{\,}$ would have a definable least element u. But then u would be invariant, and one can use the homogeneity of B to show that $u \in (P\omega)\hat{\,}$ in $V^{(B)}$, a contradiction.

Now for the formal details. Put $c = [\![\text{WO}_\phi]\!]^B$ and write S for $P\hat\omega - (P\omega)\hat{\,}$ in $V^{(B)}$. Then, by 3.7 and 3.6, either $c = 0$ or $c = 1$; we have to prove the former. Suppose, for contradiction's sake, that $c = 1$. By 2.6 we have $V^{(B)} \models S \neq \emptyset$, so it follows that

(1) $\quad V^{(B)} \models \exists! x \in S \forall y \in S \phi(x, y)$.

Hence, by the Maximum Principle, there is $u \in V^{(B)}$ such that

(2) $\quad V^{(B)} \models u \in S \wedge \forall y \in S \phi(u, y)$.

We have $V^{(B)} \models u \subseteq \hat{\omega}$ and, by (1) and (2), for all $n \in \omega$,

$$V^{(B)} \models [\hat{n} \in u \leftrightarrow \exists x \in S[\hat{n} \in x \wedge \forall y \in S \phi(x, y)]].$$

Hence

$$[\![\hat{n} \in u]\!]^B = [\![\exists x \in S[\hat{n} \in x \wedge \forall y \in S \phi(x, y)]]\!]^B.$$

Now the r.h.s. of this equation is evidently invariant, so by 3.6 and 3.7 the l.h.s. is either 0 or 1. Put

$$v = \{n \in \omega : [\![\hat{n} \in u]\!]^B = 1\}..$$

Then $[\![\hat{n} \in \hat{v}]\!]^B = [\![\hat{n} \in u]\!]^B$, so that

$$[\![\forall x \in \hat{\omega}[x \in u \leftrightarrow x \in \hat{v}]]\!]^B = 1,$$

whence $V^{(B)} \models u = \hat{v}$, and thus $V^{(B)} \models u \in (P\omega)\hat{\ }$. But this contradicts the fact—immediate from (2)—that $V^{(B)} \models u \notin (P\omega)\hat{\ }$. Thus $c = 0$ and we are through. □

3.9. Corollary. *If* ZF *is consistent, so is*

$$\text{ZFC} + \text{GCH} + \{\neg \text{WO}_\phi : \phi(x, y) \text{ a formula}\}.$$

Proof. If $B = \text{RO}(2^\omega)$ then $|B| = 2^{\aleph_0}$ and so by 2.8 in ZFC + GCH we can prove that $V^{(B)} \models \text{GCH}$. The required result now follows easily from this, 3.7, 3.8, 1.33 and 1.19. □

Problems

3.10. (*The Boolean-valued subset defined by a formula*). Let $\psi(x)$ be any B-formula and let $u \in V^{(B)}$. Recall that in the proof of 1.35 we showed that the object $v \in V^{(B)}$ defined by $\mathrm{dom}(v) = \mathrm{dom}(u)$ and $v(x) = u(x) \wedge [\![\psi(x)]\!]$ for $x \in \mathrm{dom}(v)$ satisfies $V^{(B)} \models \forall x[x \in v \leftrightarrow x \in u \wedge \psi(x)]$. Write $v = \{x \in u : \psi(x)\}^{(B)}$. Now suppose that B is homogeneous, $\phi(x, v_1, \ldots, v_n)$ is any formula and $a, a_1, \ldots, a_n \in V$. Show that

$$V^{(B)} \models \{x \in \hat{a} : \phi(x, \hat{a}_1, \ldots, \hat{a}_n)\}^{(B)} \in (Pa)^{\widehat{}}.$$

3.11. (*Ordinal definable sets in $V^{(B)}$*). We recall (cf. Drake 1974, Ch.5) that a set u is said to be *ordinal definable*—written $\mathrm{OD}(u)$—if for some formula ϕ and ordinals $\alpha_1, \ldots, \alpha_n, u$ is the unique set such that $\phi(u, \alpha_1, \ldots, \alpha_n)$ holds. The set u is *hereditarily ordinal definable*—written $\mathrm{HOD}(u)$—if all the members of the transitive closure of $\{u\}$ are ordinal definable. We put $\mathrm{OD} = \{u : \mathrm{OD}(u)\}$ and $\mathrm{HOD} = \{u : \mathrm{HOD}(u)\}$. Recall the following facts:

(a) OD has a definable well-ordering and, for any set u, $u \subseteq \mathrm{OD}$ iff u has a definable well-ordering;

(b) $L \subseteq \mathrm{HOD} \subseteq \mathrm{OD}$.

(i) Show that $L = \mathrm{HOD}$ iff for all u, $(u \subseteq L \wedge u \in \mathrm{OD}) \to u \in L$. (Use $\in$-induction.)

(ii) Suppose that B is homogeneous. Show that $V = L \to V^{(B)} \models L = \mathrm{HOD}$. (Use (i) and 3.10).

(iii) Show that, if ZF is consistent, so is ZFC + GCH + $L = \mathrm{HOD}$ + $\mathrm{HOD} \neq V$. (Use (ii) and 3.9.)

3.12. (*Complete homomorphisms*). Let B and B' be complete Boolean algebras. Recall that a homomorphism $h: B \to B'$ is said to be *complete* if $h(\bigvee X) = \bigvee h[X]$ for any $X \subseteq B$.

(i) Let h be a complete *monomorphism* of B into B'. Define the map $\bar{h}$ on $V^{(B)}$ by recursion: for all $u \in V^{(B)}$

$$\bar{h}u = \{\langle \bar{h}x, h(u(x)) \rangle : x \in \mathrm{dom}(u)\}.$$

Show that h is an injection of $V^{(B)}$ into $V^{(B')}$ such that, for any $u, v \in V^{(B)}$, $h[\![u \in v]\!]^B = [\![\bar{h}u \in \bar{h}v]\!]^{B'}$, $h[\![u = v]\!]^B = [\![\bar{h}u = \bar{h}v]\!]^B$ and, for $x \in V$, $\bar{h}\hat{x} = \hat{x}$. (Argue inductively as in the proof of 3.3.)

Throughout the remainder of this problem take h to be a complete homomorphism of B into B'.

(ii) Define the map $\tilde{h}$ on $V^{(B)}$ by recursion: for $u \in V^{(B)}$

$$\tilde{h}u = \{\langle \tilde{h}x, \bigvee_{B'} \{h(u(y)) : y \in \mathrm{dom}(u) \wedge \tilde{h}x = \tilde{h}y\}\rangle : x \in \mathrm{dom}(u)\}.$$

Show that (a) $\tilde{h} : V^{(B)} \to V^{(B')}$, (b) $\tilde{h}$ is onto if h is, (c) $\tilde{h} = \bar{h}$ if h is a monomorphism, (d) $\tilde{h}\hat{x} = \hat{x}$ for $x \in V$, (e) $h[\![u \in v]\!]^B = [\![\tilde{h}u = \tilde{h}v]\!]^{B'}$, $h[\![u = v]\!]^B = [\![\tilde{h}u = \tilde{h}v]\!]^{B'}$ for $u, v \in V^{(B)}$. (Argue inductively.)

(iii) If h' is a complete homomorphism of B' into a complete Boolean algebra B'', show that $(h' \circ h)^\sim = \tilde{h}' \circ \tilde{h}$. (Argue inductively.)

(iv) Show that for any Σ_1-formula $\phi(v_1, \ldots, v_n)$ and any $x_1, \ldots, x_n \in V^{(B)}$

$$h[\![\phi(x_1, \ldots, x_n)]\!]^B \leq [\![\phi(\tilde{h}x_1, \ldots, \tilde{h}x_n)]\!]^{B'}.$$

(v) Show that, if h is *onto*, then for any formula $\phi(v_1, \ldots, v_n)$ and any $x_1, \ldots, x_n \in V^{(B)}$,

$$h[\![\phi(x_1, \ldots, x_n)]\!]^B = [\![\phi(\tilde{h}x_1, \ldots, \tilde{h}x_n)]\!]^{B'}.$$

3.13. (*Ultrapowers as Boolean extensions*).

(i) Let I be a set and for each $i \in I$ define $\phi_i : PI \to 2$ by $\pi_i(X) = 1$ if $i \in X$, $\pi_i(X) = 0$ if $i \notin X$; then π_i is a complete homomorphism of the complete Boolean algebra PI onto 2. Let $\phi(v_1, \ldots, v_n)$ be a formula and let $x_1, \ldots, x_n \in V^{(PI)}$. Show that

$$[\![\phi(x_1, \ldots, x_n)]\!]^{PI} = \{i \in I : [\![\phi(\tilde{\pi}_i x_1, \ldots, \tilde{\pi}_i x_n)]\!]^2 = 1\}.$$

(Use 3.12(v).)

(ii) Let B be a complete Boolean algebra and let U be an *ultrafilter* in B. Define the relation $\sim_U$ on $V^{(B)}$ by $x \sim_U y \leftrightarrow [\![x = y]\!] \in U$;

3. Group Actions and Independence of AC

then $\sim_U$ is an equivalence relation on $V^{(B)}$. For each $x \in V^{(B)}$ let $x^U = \{y \in V^{(B)} : x \sim_U y\}$ be the $\sim_U$-equivalence class of x. Define the relation $\in_U$ on the class [†] $\{x^U : x \in V^{(B)}\}$ by $x^U \in_U y^U \leftrightarrow [\![x \in y]\!] \in U$. Let $V^{(B)}/U$ be the structure $\langle \{x^U : x \in V^{(B)}\}, \in_U\rangle$: this is called the *quotient* of $V^{(B)}$ by U. (For a fuller treatment of quotients, see Chapter 4.) Let $\phi(v_1, \ldots, v_n)$ be a formula and let $x_1, \ldots, x_n \in V^{(B)}$. Show that

$$V^{(B)}/U \models \phi[x_1^U, \ldots, x_n^U] \text{ iff } [\![\phi(x_1, \ldots, x_n)]\!] \in U.$$

(Induction on the complexity of ϕ, using the Maximum Principle to handle the existential case.)

(iii) Let I be a set. Define $g: V^{(2)} \to V$ by putting, for each $u \in V^{(2)}$, $g(u) = $ the unique $x \in V$ such that $V^{(PI)} \models u = \hat{x}$ (1.23(iv)). Define $h: V^{(PI)} \to V^I$, the set of all maps with domain I, by $h(x) = \{g(\tilde{\pi}_i x) : i \in I\}$ for $x \in V^{(PI)}$. Show that h is onto. (Start with $f \in V^I$; observe that $\{f^{-1}(f(i)) : i \in I\}$ is a partition of unity in PI. Hence by 1.26(i) there is $x \in V^{(PI)}$ such that $[\![x = f(i)\hat{\ }]\!]^{PI} = f^{-1}(f(i))$ for $i \in I$. Show, using 3.12 and (i), that $h(x) = f$.)

(iv) Let I be a set, let U be an ultrafilter in PI, and let V^I/U be the usual ultrapower of V by U. Define $j: V^{(PI)}/U \to V^I/U$ by $j(x^U) = h(x)/U$, the canonical image of $h(x)$ in V^I/U (and h is defined as in (iii)). Show that j is an isomorphism of $V^{(PI)}/U$ onto V^I/U. (Use (i), (ii) and (iii).) *Thus, each ultrapower of V can be obtained as a quotient of a suitable Boolean extension of V.*

The Independence of the Axiom of Choice

We turn next to the problem of establishing the relative consistency of $\neg$ AC with ZF. Now it is clear that we cannot do this by trying to falsify AC in some $V^{(B)}$ (as, e.g., we did with CH), because we know that in ZFC one can *prove* that AC is true in $V^{(B)}$. It turns out, however, that, if B is acted on by a suitable *group*, then we can falsify AC in certain *submodels* of $V^{(B)}$.

[†]Strictly speaking, each x^U is itself a (proper) class, so $\{x^U : x \in V^{(B)}\}$ is not defined. However, this annoyance can be overcome by Scott's well-known trick of replacing each x^U by the set of its members of minimum rank.

We first give a heuristic sketch of the argument.

Let G be the group of all permutations of ω and for each $n \in \omega$ let
$$G_n = \{g \in G : gn = n\}.$$
We choose a certain complete Boolean algebra B and construct a certain subclass $V^{(\Gamma)}$ of $V^{(B)}$ such that

(i) $V^{(\Gamma)}$ is a B-valued model of ZF such that $\hat{x} \in V^{(\Gamma)}$ for all $x \in V$;

(ii) G acts on $V^{(\Gamma)}$;

(iii) for each $x \in V^{(\Gamma)}$, there is a finite subset $J \subseteq \omega$ (called a *support* of x) such that $gx = x$ for every $g \in \bigcap_{n \in J} G_n$;

(iv) there is an infinite 'set of distinct reals' $\{u_n : n \in \omega\} = s$ in $V^{(\Gamma)}$ such that $gu_n = u_{gn}$ for all $g \in G$.

Then, in $V^{(\Gamma)}$, s is infinite but not Dedekind infinite, so *a fortiori* the axiom of choice fails in $V^{(\Gamma)}$. For suppose f is any map in $V^{(\Gamma)}$ of $\hat{\omega}$ into s. Then, by (iii), f has a finite support J. If f were one-one, then there would be $n \notin J$ such that $u_n \in \text{ran}(f)$. Choose $n' \notin \{n\} \cup J$ and let $g \in G$ be the permutation of ω which interchanges n and n' but leaves everything else undisturbed. If $u_n = f(\hat{m})$, then $u_{n'} = u_{gn} = gu_n = g(f(\hat{m})) = (gf)(g\hat{m}) = f(\hat{m}) = u_n$, contradicting $u_n \neq u_{n'}$. Thus there is no one-one map of $\hat{\omega}$ into s, so that s is not Dedekind infinite.

Condition (iii) implies that the members of $V^{(\Gamma)}$ have the following property: for $x \in V^{(B)}$ let $\text{stab}(x) = \{g \in G : gx = x\}$; then $\text{stab}(x) \in \Gamma$ for every $x \in V^{(\Gamma)}$, where Γ is the filter of subgroups generated by the G_n, *i.e.* $\Gamma = \{H : H \text{ a subgroup of } G \text{ and for some finite } J \subseteq \omega, \bigcap_{n \in J} G_n \subseteq H\}$. This leads us to consider an (arbitrary) *filter of subgroups* of an (arbitrary) group G. Finally, since we want G to act on $V^{(\Gamma)}$, we must have $x \in V^{(\Gamma)} \to gx \in V^{(\Gamma)} \to \text{stab}(gx) \in \Gamma$. But it is easy to verify that $\text{stab}(gx) = g\text{stab}(x)g^{-1}$, so we shall want Γ to satisfy $H \in \Gamma \to gHg^{-1} \in \Gamma$ for $g \in G$. Under these conditions Γ is said to be *normal*.

We now turn to the formal development. Let G be a group acting on the complete Boolean algebra B, and let Γ be a *filter of subgroups* of G.

3. Group Actions and Independence of AC

That is, Γ is a non-empty set of subgroups of G such that

$$H, K \in \Gamma \to H \cap K \in \Gamma$$

and

$$H \in \Gamma \text{ and } H \subseteq K, K \text{ a subgroup of } G \to K \in \Gamma.$$

Γ is called a *normal* filter if

$$g \in G \text{ and } H \in \Gamma \to gHg^{-1} \in \Gamma.$$

By 3.3, G acts on $V^{(B)}$; for each $x \in V^{(B)}$ we define the *stabilizer* of x by

$$\text{stab}(x) = \{g \in G : gx = x\}.$$

It is easy to verify that $\text{stab}(x)$ is subgroup of G. We define (by analogy with (1.4)) the sets $V_\alpha^{(\Gamma)}$ recursively as follows:

$$V_\alpha^{(\Gamma)} = \{x : \text{Fun}(x) \wedge \text{ran}(x) \subseteq B \wedge \text{stab}(x) \in \Gamma \wedge \exists \xi < \alpha [\text{dom}(x) \subseteq V_\xi^{(\Gamma)}]\}.$$

We put

$$V^{(\Gamma)} = \{x : \exists \alpha (x \in V_\alpha^{(\Gamma)})\}.$$

It is now easy to verify that (cf. (1.6)):

$$x \in V^{(\Gamma)} \leftrightarrow \text{Fun}(x) \wedge \text{ran}(x) \subseteq B \wedge \text{dom}(x) \subseteq V^{(\Gamma)} \wedge \text{stab}(x) \in \Gamma,$$

and that

$$V^{(\Gamma)} \subseteq V^{(B)}.$$

For $u, v \in V^{(\Gamma)}$ we define $[\![u \in v]\!]^\Gamma$ and $[\![u = v]\!]^\Gamma$ recursively as we did $[\![u \in v]\!]^B$ and $[\![u = v]\!]^B$, i.e.

$$[\![u \in v]\!]^\Gamma = \bigvee_{x \in \text{dom}(v)} [v(x) \wedge [\![x = u]\!]^\Gamma],$$

$$[\![u = v]\!]^\Gamma = \bigwedge_{x \in \text{dom}(u)} [u(x) \Rightarrow [\![x \in v]\!]^\Gamma] \wedge \bigwedge_{y \in \text{dom}(v)} [v(y) \Rightarrow [\![x \in u]\!]^\Gamma].$$

It is then easily proved by induction that $[\![u \in v]\!]^\Gamma = [\![u \in v]\!]^B$, $[\![u = v]\!]^\Gamma = [\![u = v]\!]^B$, and so $[\![. \in .]\!]^\Gamma$, $[\![. = .]\!]^\Gamma$ turn $V^{(\Gamma)}$ into a B-valued structure. So if $\mathcal{L}^{(\Gamma)}$ is the language for $V^{(\Gamma)}$, i.e. the result of expunging from $\mathcal{L}^{(B)}$ all constant symbols not denoting elements of $V^{(\Gamma)}$, the Boolean-value $[\![\sigma]\!]^\Gamma$ in $V^{(\Gamma)}$ of any $\mathcal{L}^{(\Gamma)}$-sentence σ is defined recursively by

$$[\![\sigma \wedge \tau]\!]^\Gamma = [\![\sigma]\!]^\Gamma \wedge [\![\tau]\!]^\Gamma$$
$$[\![\neg \sigma]\!]^\Gamma = ([\![\sigma]\!]^\Gamma)^*$$
$$[\![\exists x \phi(x)]\!]^\Gamma = \bigvee_{u \in V^{(\Gamma)}} [\![\phi(u)]\!]^\Gamma.$$

We shall need some technical facts about $V^{(\Gamma)}$.

3.14. Lemma. *For every* $x \in V$,

$$\hat{x} \in V^{(\Gamma)}.$$

Proof. By induction on $\in$. Suppose $\hat{y} \in V^{(\Gamma)}$ for every $y \in x$. Then $\text{dom}(\hat{x}) = \{\hat{y} : y \in x\} \subseteq V^{(\Gamma)}$ and by 3.3 we have $g\hat{x} = \hat{x}$ for very $g \in G$, whence $\text{stab}(x) = G \in \Gamma$. Hence $\hat{x} \in V^{(\Gamma)}$. $\square$

From now on we assume that Γ *is a normal filter of subgroups of* G.

3.15. Lemma. G *acts on* $V^{(\Gamma)}$.

Proof. One first shows by induction on $y \in \text{dom}(x)$ that for any $g \in G$, the map $x \mapsto gx$ carries $V^{(\Gamma)}$ into $V^{(\Gamma)}$. Suppose then that $x \in V^{(\Gamma)}$ and $gy \in V^{(\Gamma)}$ for all $y \in \text{dom}(x)$. Then $\text{dom}(gx) = \{gy : y \in \text{dom}(x)\} \subseteq V^{(\Gamma)}$. Also, it is readily verified that $\text{stab}(gx) = g\text{stab}(x)g^{-1}$, and therefore $\text{stab}(gx) \in \Gamma$ by the normality of Γ. Hence $gx \in V^{(\Gamma)}$, completing the induction step.

Finally, since G acts on $V^{(B)}$, we have for any $g \in G$, $u, v \in V^{(\Gamma)}$,

$$g.[\![u \in v]\!]^\Gamma = g.[\![u \in v]\!]^B = [\![gu \in gv]\!]^B = [\![gu \in gv]\!]^\Gamma$$

and similarly for $u = v$. Therefore G acts on $V^{(\Gamma)}$ as claimed. $\square$

3. Group Actions and Independence of AC

From 3.15 and 3.2 it follows that, for every formula $\phi(v_1,\ldots,v_n)$, all $x_1,\ldots,x_n \in V^{(\Gamma)}$ and all $g \in G$,

$$g.[\![\phi(x_1,\ldots,x_n)]\!]^\Gamma = [\![\phi(gx_1,\ldots,gx_n)]\!]^\Gamma. \tag{3.16}$$

If P is a basis for B, we define, for any $p \in P$ and any $\mathcal{L}^{(\Gamma)}$-sentence σ, p Γ-*forces* σ by

$$p \Vdash_\Gamma \sigma \leftrightarrow p \leq [\![\sigma]\!]^\Gamma.$$

It follows immediately from (3.16) that for any formula $\phi(v_1,\ldots,v_n)$, any $x_1,\ldots,x_n \in V^{(\Gamma)}$ and any $p \in P$, $g \in G$ for which $gp \in P$,

$$p \Vdash_\Gamma \phi(x_1,\ldots,x_n) \to gp \Vdash_\Gamma \phi(gx_1,\ldots,gx_n). \tag{3.17}$$

We define the notions of *truth* and *validity* in $V^{(\Gamma)}$ in the same way as we defined those notions in $V^{(B)}$, e.g. an $\mathcal{L}^{(\Gamma)}$-sentence σ is *true* in $V^{(\Gamma)}$ (and we write $V^{(\Gamma)} \models \sigma$) if $[\![\sigma]\!]^\Gamma = 1$.

The same argument as that used in the proof of 1.17 establishes

3.18. Theorem. 1.17 *continues to hold when 'B' is replaced by 'Γ'.* □

We can now show that $V^{(\Gamma)}$ is a Boolean-valued model of ZF.

3.19. Theorem. *All the axioms—and hence all the theorems—of* ZF *are true in* $V^{(\Gamma)}$.

Proof. The axioms of extensionality and regularity go through as in 1.34 and 1.42. As for the remaining axioms (with the exception of choice!), the same proofs as those given in 1.35–1.41 work, except that now one must verify that the object v required to exist by the axiom in question has its stabilizer stab(v) in Γ. We do this in detail for the axiom scheme of separation, confining ourselves to brief hints in the case of other axioms.

Separation. Let $\psi(x,v_1,\ldots,v_n)$ be a formula, and let $u,a_1,\ldots,a_n \in V^{(\Gamma)}$ (thus the $a_1,\ldots,a_n$ are regarded as *parameters*). Define $v \in V^{(B)}$ by dom(v) = dom(u) and

$$v(x) = u(x) \wedge [\![\psi(x,a_1,\ldots,a_n)]\!]^\Gamma.$$

for $x \in \mathrm{dom}(v)$. It now suffices to show that $v \in V^{(\Gamma)}$, for then one readily verifies as in 1.35 that

$$V^{(\Gamma)} \models \forall x[x \in v \leftrightarrow x \in u \land \psi(x, a_1, \ldots, a_n)].$$

Since $\mathrm{dom}(v) = \mathrm{dom}(u) \subseteq V^{(\Gamma)}$, to show that $v \in V^{(\Gamma)}$ it is enough to prove that $\mathrm{stab}(v) \in \Gamma$. And since $\mathrm{stab}(u), \mathrm{stab}(a_1)\ldots, \mathrm{stab}(a_n)$ are all in Γ and Γ is a filter, it will be enough to show that

$$A = \mathrm{stab}(u) \cap \mathrm{stab}(a_1) \cap \ldots \cap \mathrm{stab}(a_n) \subseteq \mathrm{stab}(v). \qquad (*)$$

If $g \in A$, then $\mathrm{dom}(gv) = \{gx : x \in \mathrm{dom}(v)\} = \{gx : x \in \mathrm{dom}(u)\} = \mathrm{dom}(gu) = \mathrm{dom}(u) = \mathrm{dom}(v)$. Also, if $x \in \mathrm{dom}(v)$, then $x = gy$ with $y \in \mathrm{dom}(u)$ so that

$$\begin{aligned}(gv)(x) &= (gv)(gy) \\ &= g.v(y) \\ &= g.u(y) \land [\![\psi(gy, ga_1, \ldots, ga_n)]\!]^\Gamma \quad \text{(by (3.16))} \\ &= (gu)(gy) \land [\![\psi(x, a_1, \ldots, a_n)]\!]^\Gamma \\ &= u(x) \land [\![\psi(x, a_1, \ldots, a_n)]\!]^\Gamma \\ &= v(x).\end{aligned}$$

Hence $gv = v$ and $g \in \mathrm{stab}(v)$. This proves $(*)$ and the result in question.

Replacement. In our original proof of the truth of this axiom in $V^{(B)}$ (1.36) we used a set of the form $V_\alpha^{(B)} \times \{1\}$ to include the range of the 'function' defined by $\phi(x, y)$ on u. The same proof works here with $V_\alpha^{(B)} \times \{1\}$ replaced by $V_\alpha^{(\Gamma)} \times \{1\}$.

Union. Given $u \in V^{(\Gamma)}$, define v by $\mathrm{dom}(v) = \bigcup \{\mathrm{dom}(y) : y \in \mathrm{dom}(u)\}$ and $v(x) = [\![\exists y \in u[x \in y]]\!]^\Gamma$. One then verifies that $\mathrm{stab}(u) \subseteq \mathrm{stab}(v)$, so that $v \in V^{(\Gamma)}$. As in the original verification of the truth of the union axiom in $V^{(B)}$ (1.37) one shows that

$$V^{(\Gamma)} \models \forall x[x \in v \leftrightarrow \exists y \in u[x \in y]],$$

so that the axiom is true in $V^{(\Gamma)}$ as well.

Power Set. For $u \in V^{(\Gamma)}$ define V by $\mathrm{dom}(v) = B^{\mathrm{dom}(u)} \cap V^{(\Gamma)}$ and $v(x) = [\![x \subseteq u]\!]^\Gamma$ for $x \in \mathrm{dom}(v)$. One can now show that $\mathrm{stab}(u) \subseteq \mathrm{stab}(v)$, so

3. Group Actions and Independence of AC

that $v \in V^{(\Gamma)}$. As in the original proof of the truth of the power set axiom in $V^{(B)}$ (1.38), one verifies that

$$V^{(\Gamma)} \models \forall x[x \in v \leftrightarrow x \subseteq u],$$

showing that the axiom is true in $V^{(\Gamma)}$ as well.

Infinity. We know that $\hat{\omega} \in V^{(\Gamma)}$ (3.14), so the argument here is the same as in 1.41. □

We now select specific B, G and Γ in such a way that we have $V^{(\Gamma)} \models \neg \mathrm{AC}$. It is clear that, if we can prove the existence of such B, G and Γ in ZFC, then in view of 3.18 and 3.19, the same argument as we used to prove 1.19 implies that, if ZF is consistent, so is $\mathrm{ZF} + \neg \mathrm{AC}$.

Let X be the product space $2^{\omega \times \omega}$ and let $B = \mathrm{RO}(2^{\omega \times \omega})$. Let G be the group of all permutations of ω. G can be made to act on B in the following way. Each $g \in G$ induces a homeomorphism g^* of X onto itself via

$$(g^* f)(m, n) = f(m, gn)$$

for $f \in X$ and $m, n \in \omega$. We define the action $\langle g, b \rangle \mapsto gb$ of G on B by

$$\begin{aligned} gb &= g^{*-1}[b] \\ &= \{f \in X : g^* f \in b\}. \end{aligned}$$

(It is readily checked that this does indeed define an action.)

For each $n \in \omega$ let $G_n = \{g \in G : gn = n\}$; clearly this is a subgroup of G. Let Γ be the filter of subgroups generated by the G_n. That is, if for each finite subset $J \subseteq \omega$ we write

$$G_J = \bigcap_{n \in J} G_n$$

then Γ is the set of all subgroups H of G such that $G_J \subseteq H$ for some finite $J \subseteq \omega$. It is readily verified that Γ is normal.

Recalling that $P = C(\omega \times \omega, 2)$ is a basis for B, we now prove

3.20. Lemma. *If $p \in P$, J is a finite subset of ω and $n \notin J$, then there is $g \in G_J$ such that $p \wedge gp \neq 0$ and $gn \neq n$.*

Proof. Take $n' \notin J \cup \{n\}$ so that $\langle m, n' \rangle \notin \text{dom}(p)$ for any m (possible, since J and $\text{dom}(p)$ are finite!) and let $g \in G$ be the permutation of ω which interchanges n and n' but leaves everything else undisturbed. Then certainly $g \in G_J$ and $gn \neq n$. To see that $p \wedge gp \neq 0$, recall that p is identified with the element

$$N(p) = \{f \in 2^{\omega \times \omega} : p \subseteq f\}$$

of B and observe that

$$\begin{aligned} g.N(p) &= \{f \in 2^{\omega \times \omega} : p \subseteq g^*f\} \\ &= \{f \in 2^{\omega \times \omega} : \langle i, j \rangle \in \text{dom}(p) \to f\langle i, gj \rangle = p\langle i, j \rangle\}. \end{aligned}$$

Let $i_1, \ldots, i_k$ be a list of the i such that $\langle i, n \rangle \in \text{dom}(p)$. Then

$$\begin{aligned} p \wedge gp &= N(p) \cap g.N(p) \\ &= \{f \in 2^{\omega \times \omega} : p \subseteq f \text{ and } f\langle i_j, n' \rangle = p\langle i_j, n \rangle \text{ for } j = 1, \ldots, k\} \\ &\neq \emptyset, \end{aligned}$$

since $\langle i_j, n' \rangle \notin \text{dom}(p)$ for $j = 1, \ldots, k$. □

Now we can prove

3.21. Theorem. *With B, G, and Γ as above, we have $V^{(\Gamma)} \models$ 'there is an infinite Dedekind finite subset of $P\hat{\omega}$' and so, a fortiori,*

$$V^{(\Gamma)} \models \neg \text{AC}.$$

Proof. We write $[\![\sigma]\!]$ for $[\![\sigma]\!]^\Gamma$ and $\Vdash$ for $\Vdash_\Gamma$ throughout. For each $n \in \omega$ define $u_n \in B^{\text{dom}(\hat{\omega})}$ by

$$u_n(\hat{m}) = \{h \in 2^{\omega \times \omega} : h\langle m, n \rangle = 1\}.$$

The usual calculation shows that

$$V^{(B)} \models u_n \subseteq \hat{\omega}$$

for all $n \in \omega$. Moreover for all $g \in G$ and for all $n \in \omega$,

3. Group Actions and Independence of AC

(1) $\quad gu_n = u_{gn}$.

For clearly we have $\text{dom}(gu_n) = \text{dom}(u_{gn})$. Also, for $m \in \omega$,

$$\begin{aligned}(gu_n)\hat{m} &= (gu_n)(g\hat{m}) \\ &= g.u_n(\hat{m}) \\ &= g^{*-1}[\{h \in 2^{\omega \times \omega} : h\langle m, n\rangle = 1\}] \\ &= \{h \in 2^{\omega \times \omega} : g^*h\langle m, n\rangle = 1\} \\ &= \{h \in 2^{\omega \times \omega} : h\langle m, gn\rangle = 1\} \\ &= u_{gn}(\hat{m}),\end{aligned}$$

and (1) follows.

(1) immediately gives $G_n \subseteq \text{stab}(u_n)$, so $\text{stab}(u_n) \in \Gamma$ and therefore $u_n \in V^{(\Gamma)}$. The argument in the proof of 2.12 gives

(2) $\quad V^{(\Gamma)} \models u_n \neq u_{n'}$, for $n \neq n'$.

Now put

$$s = \{u_n : n \in \omega\} \times \{1\};$$

it is then easy to see that $gs = s$ for any $g \in G$, so $s \in V^{(\Gamma)}$. Moreover, it is not hard to verify that $V^{(\Gamma)} \models s \subseteq P\hat{\omega}$. It now follows from (2) that

$$V^{(\Gamma)} \models s \text{ is infinite}.$$

We claim that

$$V^{(\Gamma)} \models s \text{ is not Dedekind infinite},$$

which will prove the theorem. To establish the claim, it suffices to show that, for each $f \in V^{(\Gamma)}$,

$$[\![\text{Fun}(f) \land f \text{ is one-one} \land \text{dom}(f) = \hat{\omega} \land \text{ran}(f) \subseteq s]\!] = 0.$$

Suppose not; then there is $p_0 \in P = C(\omega \times \omega, 2)$ (the basis for B) such that

$$p_0 \Vdash \text{Fun}(f) \land f \text{ is one-one} \land \text{dom}(f) = \hat{\omega} \land \text{ran}(f) \subseteq s.$$

We shall find $q \leq p_0$ such that

$$q \Vdash \neg \mathrm{Fun}(f),$$

which will yield the desired contradiction.

We first observe that

(3) $\quad p \Vdash x \in s \leftrightarrow \forall q \leq p \exists r \leq q \exists n \in \omega [r \Vdash x = u_n].$

For we have

$$\begin{aligned}
p \Vdash x \in s &\leftrightarrow p \leq \bigvee_{n \in \omega} [\![x = u_n]\!] \\
&\leftrightarrow p \wedge \bigwedge_{n \in \omega} [\![x \neq u_n]\!] = 0 \\
&\leftrightarrow \forall q \leq p [q \not\leq \bigwedge_{n \in \omega} [\![x \neq u_n]\!]] \\
&\leftrightarrow \forall q \leq p \exists n \in \omega [q \not\leq [\![x \neq u_n]\!]] \\
&\leftrightarrow \forall q \leq p \exists n \in \omega \neg [q \Vdash x \neq u_n] \\
&\leftrightarrow \forall q \leq p \exists n \in \omega \exists r \leq q [r \Vdash x = u_n].
\end{aligned}$$

Now since $f \in V^{(\Gamma)}$ it has a finite support J, i.e. there is a finite subset $J \subseteq \omega$ such that $G_J \subseteq \mathrm{stab}(f)$. Let

$$J = \{n_1, \ldots, n_j\}.$$

Since $p_0 \Vdash f$ is one-one $\wedge \mathrm{Fun}(f)$, it follows that

$$p_0 \Vdash \exists x \in \hat{\omega}[f(x) \neq u_{n_1} \wedge \ldots \wedge f(x) \neq u_{n_j}]$$

so that there is $p \leq p_0$ and $m \in \omega$ such that

(4) $\quad p \Vdash f(\hat{m}) \neq u_{n_1} \wedge \ldots \wedge f(\hat{m}) \neq u_{n_j}.$

Since $p_0 \Vdash f(\hat{m}) \in s$, so that $p \Vdash f(\hat{m}) \in s$, by (3) there is $r \leq p$ and $n \in \omega$ such that

(5) $\quad r \Vdash f(\hat{m}) = u_n.$

But (4) implies

$$r \Vdash f(\hat{m}) \neq u_{n_1} \wedge \ldots \wedge f(\hat{m}) \neq u_{n_j}$$

3. Group Actions and Independence of AC

and this, together with (5), implies $n \neq n_1 \wedge \ldots \wedge n \neq n_j$, i.e. $n \notin J$. By 3.20 there is $g \in G_J$ such that $p \wedge gp = 0$ and $gn \neq n$. It follows from (5) and (3.17) that

$$gr \Vdash (gf)g\hat{m} = gu_n.$$

But this, together with (1) and the fact that $g \in G_J \subseteq \mathrm{stab}(f)$ gives

$$gr \Vdash f(\hat{m}) = u_{gn}.$$

Since $r \wedge gr \neq 0$, there is $q \in P$ such that $q \leq r$ and $q \leq gr$. Then $q \leq p_0$ and

$$q \Vdash f(\hat{m}) = u_n \wedge f(\hat{m}) = u_{gn}.$$

But since $gn \neq n$, we have, using (2), $[\![u_{gn} \neq u_n]\!] = 1$, so that $q \Vdash u_{gn} \neq u_n$. Therefore

$$q \Vdash \neg \mathrm{Fun}(f),$$

and the proof is complete. □

3.22. Corollary. *If* ZF *is consistent, so is* ZF $+ \neg$AC. □

We conclude this chapter with some remarks on the origins of the proof of Theorem 3.21. The construction of $V^{(\Gamma)}$ is in fact derived from an earlier construction, due to Fraenkel and Mostowski, which was used to show that AC is independent of a certain modified form of ZF, namely, the theory ZFA of set theory with atoms. (This is actually a weaker result because the axiom of foundation does not hold in ZFA.) To obtain ZFA from ZF, one drops the axiom of foundation, and adds an axiom asserting the existence of a non-empty set A of *atoms*, *i.e.* objects with no members, yet not identical with the empty set. The axiom of extensionality must also be suitably modified. The method of Fraenkel-Mostowski now runs roughly as follows. One first shows that the permutation group G of A acts on the universe V by $\in$-automorphisms. Then, letting Γ be a normal filter in G, one constructs $V^{(\Gamma)}$ essentially as before and shows that it is a model of ZFA. By choosing A and Γ properly, one can arrange things so that in $V^{(\Gamma)}$ the set A is not well-orderable. This will be so essentially because the presence in $V^{(\Gamma)}$ of so many automorphisms permuting the elements of A will make these elements effectively indiscernible in $V^{(\Gamma)}$ and thereby render it impossible to choose a 'first element' of A. (For more on Fraenkel-Mostowski models, see Felgner 1971 or Jech 1973.)

Our method for proving the independence of AC from ZF combines the Fraenkel-Mostowski technique with that of Boolean-valued models. In essence, we found a set of 'reals' in $V^{(\Gamma)}$ (the set $\{u_n : n \in \omega\}$) which behaves very much like a set of atoms, and then argued à la Fraenkel-Mostowski. It turns out that many Fraenkel-Mostowski proofs of independence from ZFA can be converted in this way into proofs of independence from ZF, e.g. those of 'every set can be linearly ordered', 'every countable set of pairs has a choice function', 'every vector space has a basis'.

Chapter 4

Generic Ultrafilters and Transitive Models of ZFC

In this chapter we shall examine what happens when we replace V by a (transitive) model M of ZFC such that $M \in V$, and perform the construction of $V^{(B)}$ and $[\![\cdot]\!]^B$ *inside* M. We shall see that this construction gives rise to models of ZFC in which one can falsify the various set-theoretic assertions whose formal independence of ZFC was established in earlier chapters. In this way Boolean-valued set theory can be transformed into a valuable model-theoretic tool.

Now let M be a transitive $\in$-model of ZFC and let $B \in M$ be a complete Boolean algebra in the sense of M. That is,

$$M \models B \text{ is a complete Boolean algebra}.$$

In particular, if $X \in PB \cap M$, then $\bigvee X$ and $\bigwedge X$ exist and are in M. (Notice that $PB \cap M$ is the power set of B formed in M, $P^{(M)}(B)$.) Moreover, since the predicate 'B is a Boolean algebra' is a restricted formula, it follows that B is a Boolean algebra (but not necessarily complete!) 'from the outside' as well.

Under these conditions we can relativize all the notions and constructions of Chapter 1 to M. We write $M^{(B)}$ for $(V^{(B)})^{(M)}$ and $\mathcal{L}_M^{(B)}$ for $(\mathcal{L}^{(B)})^{(M)}$. $M^{(B)}$ is called the *B-extension* of M: all the results proved in Chapter 1 for $V^{(B)}$ hold, *mutatis mutandis*, for $M^{(B)}$. We also obtain a Boolean truth value $([\![\sigma]\!]^B)^{(M)} \in B$ for each $\mathcal{L}_M^{(B)}$-sentence σ: to simplify the notation we agree to write $[\![\sigma]\!]$ for $([\![\sigma]\!]^B)^{(M)}$. Writing $M^{(B)} \models \sigma$ for $[\![\sigma]\!] = 1$, we see from 1.33 that $M^{(B)} \models \sigma$ whenever σ is an axiom of ZFC, so that $M^{(B)}$ may be thought of as a Boolean-valued *model* of ZFC. Finally, by analogy with 1.22, we obtain a map $\hat{\ }: M \to M^B$ such that, for each $x \in M$,

$$\hat{x} = \{\langle \hat{y}, 1\rangle : y \in x\}.$$

Now suppose that U is an arbitrary but fixed *ultrafilter* in B. In general, U is *not* a member of M! Define the relation $\sim_U$ on $M^{(B)}$ by putting, for $x, y \in M^{(B)}$,

$$x \sim_U y \leftrightarrow [\![x = y]\!] \in U.$$

It is easy to verify that $\sim_U$ is an *equivalence relation* on $M^{(B)}$. For $x \in M^{(B)}$, write x^U for the $\sim_U$-class of x, i.e.

$$x^U = \{y \in M^{(B)} : x \sim_U y\},$$

and define the relation $\in_U$ on the set of all $\sim_U$-classes by

$$x^U \in_U y^U \leftrightarrow [\![x \in y]\!] \in U.$$

Now define the *quotient* of $M^{(B)}$ by U to be the structure

$$M^{(B)}/U = \langle \{x^U : x \in M^{(B)}\}, \in_U\rangle.$$

4.1. Theorem. *For any formula $\phi(v_1, \ldots, v_n)$ and any $x_1, \ldots, x_n \in M^{(B)}$,*

$$M^{(B)}/U \models \phi[x_1^U, \ldots, x_n^U] \leftrightarrow [\![\phi(x_1, \ldots, x_n)]\!] \in U.$$

Proof. Induction on the complexity of ϕ, using the Maximum Principle to handle the existential case. We omit the straightforward details. □

4.2. Corollary. $M^{(B)}/U$ *is a model of ZFC. More generally, for any sentence σ, if $M^{(B)} \models \sigma$, then $M^{(B)}/U \models \sigma$.* □

Let $S \subseteq P^{(M)}(B) = PB \cap M$. Recall (Chapter 0) that U is said to be S-complete if, for all $X \in S$,

$$\bigvee X \in U \to X \cap U \neq \emptyset$$

4. Generic Ultrafilters and Models of ZFC

(the reverse implication holding trivially). A $P^{(M)}(B)$-complete ultrafilter is called *M-generic*.

A partition of unity $\{a_i : i \in I\}$ in B is called an *M-partition of unity in B* if $\langle a_i : i \in I \rangle \in M$. We have the following simple characterization of M-genericity in terms of this notion:

4.3. Lemma. *The following conditions are equivalent:*

(i) U *is M-generic;*

(ii) *for any M-partition of unity $\{a_i : i \in I\}$ in B, there is $i \in I$ such that $a_i \in U$.*

Proof. (i) $\to$ (ii) is clear. Conversely, assume (ii) and let $A \in P^{(M)}(B)$. Since the axiom of choice holds in M, there is an ordinal $\alpha \in M$ such that $A \cup \{(\bigvee A)^*\} = \{a_\xi : \xi < \alpha\}$. Now put $b_\xi = a_\xi - \bigvee_{\eta < \xi} a_\eta$ for $\xi < \alpha$. Then $\{b_\xi : \xi < \alpha\}$ is an M-partition of unity in B and so by (ii) there is $\xi < \alpha$ such that $b_\xi \in U$. It is now easy to see that $\bigvee A \in U \to U \cap A \neq \emptyset$. Hence U is M-generic. □

An element $\alpha \in M^{(B)}/U$ is called an *ordinal in* $M^{(B)}/U$ if $M^{(B)}/U \models \mathrm{Ord}[\alpha]$. It follows immediately from 4.1 that if α is an ordinal in M then $\hat{\alpha}^U$ is an ordinal in $M^{(B)}/U$; elements of the latter form are called *standard ordinals in* $M^{(B)}/U$.

Let $\mathrm{ORD}^{(M)}$ be the set of all ordinals in M (thus $\mathrm{ORD}^{(M)} = \mathrm{ORD} \cap M$). Then for $x \in M^{(B)}$ the set

$$\{[\![x = \hat{\alpha}]\!] : \alpha \in \mathrm{ORD}^{(M)}\} \qquad (4.4)$$

is a subset of B which is definable in M (from the parameter x) and is therefore a member of M. Let S_1 be the subfamily of $P^{(M)}(B)$ consisting of all sets of the form (4.4) for $x \in M^{(B)}$. By 1.44 we have, for $x \in M^{(B)}$,

$$[\![\mathrm{Ord}(x)]\!] = \bigvee_{\alpha \in \mathrm{ORD}^{(M)}} [\![x = \hat{\alpha}]\!]. \qquad (4.5)$$

We use this in the proof of

4.6. Theorem. *The following conditions are equivalent:*

(i) U is S_1-complete;

(ii) all ordinals in $M^{(B)}/U$ are standard;

(iii) U is M-generic.

Proof. (i) $\to$ (iii). Assume (i) and let $\{a_\xi : \xi < \alpha\}$ be an M-partition of unity in B, where $\alpha \in \text{ORD}^{(M)}$. (Since the axiom of choice holds in M there is no loss of generality in assuming the partition of unity to be indexed by an ordinal α.) By 1.26(i), there is $x \in M^{(B)}$ such that $a_\xi = [\![x = \hat{\xi}]\!]$ for $\xi < \alpha$. We have

$$\bigvee_{\xi \in \text{ORD}^{(M)}} [\![x = \hat{\xi}]\!] \geq \bigvee_{\xi < \alpha} [\![x = \hat{\xi}]\!] = \bigvee_{\xi < \alpha} a_\xi = 1 \in U,$$

so that, by (i), there is $\eta \in \text{ORD}^{(M)}$ such that $[\![x = \hat{\eta}]\!] \in U$. If $\eta \geq \alpha$, then

$$[\![x = \hat{\eta}]\!] = [\![x = \hat{\eta}]\!] \wedge 1 = [\![x = \hat{\eta}]\!] \wedge \bigvee_{\xi < \alpha} [\![x = \hat{\xi}]\!] = 0 \notin U,$$

so that we must have $\eta < \alpha$. Hence $a_\eta = [\![x = \hat{\eta}]\!] \in U$ and (iii) follows by Lemma 4.3.

(iii) $\to$ (ii). Assume (iii); then, using (4.5) and 4.1,

$$M^{(B)}/U \models \text{Ord}[x^U] \leftrightarrow [\![\text{Ord}(x)]\!] \in U$$
$$\leftrightarrow \bigvee_{\alpha \in \text{ORD}^{(M)}} [\![x = \hat{\alpha}]\!] \in U$$
$$\leftrightarrow [\![x = \hat{\alpha}]\!] \in U \text{ for some } \alpha \in \text{ORD}^{(M)}$$
$$\leftrightarrow x^U = \hat{\alpha}^U \text{ for some } \alpha \in \text{ORD}^{(M)}.$$

(ii) $\to$ (i). Assume (ii); then if $\bigvee_{\alpha \in \text{ORD}^{(M)}} [\![x = \hat{\alpha}]\!] \in U$, we have $[\![\text{Ord}(x)]\!] \in U$ by (4.5), so $M^{(B)}/U \models \text{Ord}[x^U]$ by 4.1. Hence (ii) gives $x^U = \hat{\alpha}^U$ for some $\alpha \in \text{ORD}^{(M)}$, whence $[\![x = \hat{\alpha}]\!] \in U$. □

4.7. Corollary. [†] *If U is M-generic, then $\in_U$ is a well-founded relation.*

Proof. If U is M-generic, then 4.6 implies that the map $\alpha \mapsto \hat{\alpha}^U$ sends the well-ordered set $\text{ORD}^{(M)}$ onto the set of ordinals in $M^{(B)}/U$. This map is

---
[†] The converse of 4.7 fails: see 4.32.

easily seen to be order-preserving (with respect to $\in_U$) and it follows that the ordinals of $M^{(B)}/U$ are well-ordered by $\in_U$. The usual rank argument now shows that $\in_U$ is well-founded on $M^{(B)}/U$: if not, then there would be an infinite descending $\in_U$-sequence $\ldots x_2 \in_U x_1 \in_U x_0$; if ρ is the rank function in $M^{(B)}/U$, then $\ldots, \rho(x_2), \rho(x_1), \rho(x_0)$ would be an infinite descending sequence of ordinals in $M^{(B)}/U$, contradicting the fact that these are well-ordered. □

Suppose now that $\in_U$ is a well-founded relation. Then, by Mostowski's collapsing lemma, $M^{(B)}/U$ can be collapsed to a unique transitive $\in$-structure $M[U]$ via the map h defined recursively on $\in_U$ by

$$h(x^U) = \{h(y^U) : y^U \in_U x^U\} = \{h(y^U) : [\![y \in x]\!] \in U\}. \tag{4.8}$$

Thus $h \colon M^{(B)}/U \to M[U]$ is a bijection satisfying

$$x^U \in_U y^U \leftrightarrow h(x^U) \in h(y^U).$$

We can now define a map i of $M^{(B)}$ onto $M[U]$ by putting

$$i(x) = h(x^U) \tag{4.9}$$

for $x \in M^{(B)}$. By (4.8) we have, for $x \in M^{(B)}$,

$$i(x) = \{i(y) : [\![y \in x]\!] \in U\}. \tag{4.10}$$

The map i—which we sometimes write as i_U—is called the *canonical map of $M^{(B)}$ onto $M[U]$*.

4.11. Lemma. *For any formula $\phi(v_1, \ldots, v_n)$ and any $x_1, \ldots, x_n \in M^{(B)}$,*

$$M[U] \models \phi[i(x_1), \ldots, i(x_n)] \leftrightarrow [\![\phi(x_1, \ldots, x_n)]\!] \in U.$$

Proof. Immediate from Theorem 4.1 and the fact that h is an isomorphism of $M^{(B)}/U$ onto $M[U]$. □

Now define $j \colon M \to M[U]$ by

$$j(x) = i(\hat{x}) \tag{4.12}$$

for $x \in M$. We see immediately from 1.23 that j is a one-one map satisfying $x \in y \leftrightarrow j(x) \in j(y)$; that is, j is an $\in$-monomorphism of M into $M[U]$. The situation can be depicted as follows:

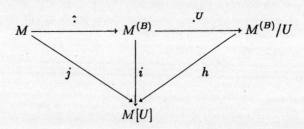

If $x \in M^{(B)}, y \in M$, the set

$$\{[\![x = \hat{z}]\!] : z \in y\} \tag{4.13}$$

is a subset of B which is definable in M (from the parameters x, y) and is accordingly a member of M. Let S_2 be the subfamily of $P^{(M)}(B)$ consisting of all sets of the form (4.13) for $x \in M^{(B)}, y \in M$. We recall from 1.23(i) that, for $x \in M^{(B)}, y \in M$,

$$[\![x \in \hat{y}]\!] = \bigvee_{z \in y} [\![x = \hat{z}]\!]. \tag{4.14}$$

We use this in the proof of

4.15. Theorem. *The following conditions are equivalent:*

(i) U is S_2-complete,

(ii) $\in_U$ is well-founded and j is the identity on M;

(iii) $\in_U$ is well-founded and $j[M]$ is transitive;

(iv) U is M-generic.

Proof. The equivalence of (ii) and (iii) follows easily from the transitivity of M and the fact that, by construction, j is an $\in$-isomorphism of M onto $j[M]$.

(i) → (iv). Assume (i), and let $\{a_i : i \in I\}$ be an M-partition of unity in B. By 1.26(i) there is $x \in M^{(B)}$ such that $a_i = [\![x = \hat{a}_i]\!]$ for $i \in I$. Putting $\{a_i : i \in I\} = y \in M$, we have

$$\bigvee_{z \in y} [\![x = \hat{z}]\!] = \bigvee_{i \in I} [\![x = \hat{a}_i]\!] = \bigvee_{i \in I} a_i = 1 \in U.$$

Therefore, by (i), there is $z \in y$ such that $[\![x = \hat{z}]\!] \in U$. But $z = a_i$ for some $i \in I$, whence $a_i = [\![x = \hat{a}_i]\!] \in U$. (iv) now follows from Lemma 4.3.

(iv) → (iii). Assume (iv). Then $\in_U$ is well-founded by 4.7. Also, if $y \in M$ and $x \in j(y)$, then, since $M[U]$ is transitive, there is $x' \in M^{(B)}$ such that $x = i(x')$. Thus $i(x') \in j(y) = i(\hat{y})$, so that, by (4.10), $[\![x' \in \hat{y}]\!] \in U$. It follows now from (iv) and (4.14) that there is $z \in y$, hence $z \in M$, such that $[\![x' = \hat{z}]\!] \in U$, whence $x = j(z) \in j[M]$, and (iii) follows.

(iii) → (i). Assume (iii). If $x \in M^{(B)}$, $y \in M$, then

$$\bigvee_{z \in y} [\![x = \hat{z}]\!] \in U \to [\![x \in \hat{y}]\!] \in U \qquad \text{(by (4.14))}$$

$$\begin{aligned}
&\to x^U \in_U \hat{y}^U \\
&\to h(x^U) \in h(\hat{y}^U) \\
&\to i(x) \in i(\hat{y}) = j(y) \\
&\to i(x) = j(z) \text{ for some } z \in M \qquad \text{(by (iii))} \\
&\to j(z) \in j(y) \\
&\to z \in y.
\end{aligned}$$

Hence $i(x) = j(z) = i(\hat{z})$, so by definition of i (4.9), $[\![x = \hat{z}]\!] \in U$. (i) follows. □

4.16. Corollary. *If U is M-generic, then $M \subseteq M[U]$.* □

Next, we define $U_* \in M^{(B)}$ by $\mathrm{dom}(U_*) = \{\hat{x} : x \in B\} = \mathrm{dom}(\hat{B})$ and $U_*(\hat{x}) = x$ for $x \in B$. Notice that U_* does *not* depend on U! We have for $x \in M^{(B)}$,

$$[\![x \in U_*]\!] = \bigvee_{y \in B} [y \wedge [\![x = \hat{y}]\!]], \qquad (4.17)$$

and, for $x \in B$,

$$[\![\hat{x} \in U_*]\!] = x. \tag{4.18}$$

In particular, if σ is any B-sentence, we have $[\![[\![\sigma]\!]\hat{\ } \in U_*]\!] = [\![\sigma]\!]$, i.e.

$$M^{(B)} \models [[\![\sigma]\!]\hat{\ } \in U_* \leftrightarrow \sigma].$$

In other words, U_* represents the truth values of "true" sentences in $M^{(B)}$.

Let S_3 be the subfamily of $P^{(M)}(B)$ consisting of all subsets of B of the form $\{y \wedge [\![x = \hat{y}]\!] : y \in B\}$ for $x \in M^{(B)}$. Then we have

4.19. Theorem. *The following conditions are equivalent:*

(i) U is S_3-complete;

(ii) $\in_U$ is well-founded, j is the identity on U, and $i(U_*) = U$;

(iii) U is M-generic.

Proof. (i) $\to$ (iii). Assume (i). Let $\{a_i : i \in I\}$ be an M-partition of unity in B. By 1.26(i) there is $x \in M^{(B)}$ such that $a_i = [\![x = \hat{a}_i]\!]$ for $i \in I$. Hence

$$\bigvee_{y \in B} [y \wedge [\![x = \hat{y}]\!]] \geq \bigvee_{i \in I} [a_i \wedge [\![x = \hat{a}_i]\!]] = \bigvee_{i \in I} a_i = 1 \in U,$$

and so by (i) there is $y \in B$ such that $y \wedge [\![x = \hat{y}]\!] \in U$. If $y \notin \{a_i : i \in I\}$ then

$$[\![x = \hat{y}]\!] = [\![x = \hat{y}]\!] \wedge 1 = [\![x = \hat{y}]\!] \wedge \bigvee_{i \in I} [\![x = \hat{a}_i]\!] = 0 \notin U,$$

so that there must be $i \in I$ such that $y = a_i$. But then $a_i = a_i \wedge [\![x = \hat{a}_i]\!] \in U$ and so (iii) follows by Lemma 4.3.

(iii) $\to$ (ii). Assume (iii). Then by 4.15 $\in_U$ is well-founded and j is the identity on M, hence on U. We claim that, for $x \in M^{(B)}$,

$$\{i(x) : [\![x \in U_*]\!] \in U\} = \{i(\hat{y}) : y \in U\}. \tag{1}$$

4. Generic Ultrafilters and Models of ZFC

In fact this follows from the chain of equivalences

$$[\![x \in U_*]\!] \in U \leftrightarrow \bigvee_{y \in B} [y \wedge [\![x = \hat{y}]\!]] \in U \qquad \text{(by (4.17))}$$
$$\leftrightarrow \exists y \in B [y \wedge [\![x = \hat{y}]\!] \in U] \qquad \text{(by (iii))}$$
$$\leftrightarrow \exists y \in U [[\![x = \hat{y}]\!] \in U]$$
$$\leftrightarrow \exists y \in U [i(x) = i(\hat{y})].$$

Now we have

$$i(U_*) = \{i(x) : [\![x \in U_*]\!] \in U\} \qquad \text{(by (4.10))}$$
$$= \{i(\hat{y}) : y \in U\} \qquad \text{(by (1))}$$
$$= \{j(y) : y \in U\}$$
$$= \{y : y \in U\} = U,$$

and this gives (ii).

(ii) → (i). Assuming (ii) we have, using (4.10),

$$U = i(U_*) = \{i(x) : [\![x \in U_*]\!] \in U\}. \qquad (2)$$

Hence

$$\bigvee_{y \in B} [y \wedge [\![x = \hat{y}]\!]] \in U \to [\![x \in U_*]\!] \in U \qquad \text{(by (4.17))}$$
$$\to i(x) \in U \qquad \text{(by (2))}$$
$$\to i(x) = y = j(y) \text{ for some } y \in U \qquad \text{(by (ii))}$$
$$\to i(x) = i(\hat{y})$$
$$\to [\![x = \hat{y}]\!] \in U$$
$$\to y \wedge [\![x = \hat{y}]\!] \in U$$

and (i) follows. □

4.19 immediately gives

4.20. Corollary. *If U is M-generic, then $U \in M[U]$.* □

We now show that U_* is an 'M-generic ultrafilter' in the sense of $M^{(B)}$. Given a Boolean algebra A and a subset $F \subseteq PA$, we say that A is *F-complete* if for all $X \in F$, $\bigvee X$ and $\bigwedge X$ exist in A. Thus our given

Boolean algebra B is $P^{(M)}(B)$-complete. Since the predicate 'A is an F-complete Boolean algebra' is clearly a restricted formula (with parameters A and F), it follows that, using 1.23,

$$M^{(B)} \models \hat{B} \text{ is a } (P^{(M)}(B))\hat{\ }\text{-complete Boolean algebra},$$

and furthermore we have

4.21. Theorem. $M^{(B)} \models U_* \text{ is a } (P^{(M)}(B))\hat{\ }\text{-complete ultrafilter in } \hat{B}$.

Proof. Let us put $C = \hat{B}$. Then we know that

$$M^{(B)} \models C \text{ is a } (P^{(M)}(B))\hat{\ }\text{-complete Boolean algebra}.$$

Let us denote the Boolean operations in C by $\wedge_C$, $*^C$, $\bigvee_C$, etc. and the natural partial ordering in C by $\leq_C$. It is then easy to see that, for any $a, b \in B$, $A \subseteq B$, we have

$$[\![\hat{a} \wedge_C \hat{b} = (a \wedge b)\hat{\ }]\!] = [\![\hat{a}^{*C} = (a^*)\hat{\ }]\!] = 1$$
$$[\![\bigvee_C \hat{A} = (\bigvee A)\hat{\ }]\!] = 1.$$

We turn now to the properties of U_*. To prove the theorem it will be enough to verify the assertions (a)–(e) below.

(a) $[\![U_* \subseteq C \wedge \hat{0} \notin U_*]\!] = 1$. This follows in a straightforward way from (4.17) and (4.18).

(b) $[\![\forall xy \in C[x, y \in U_* \to x \wedge_C y \in U_*]]\!] = 1$. To verify this, notice that the l.h.s. is

$$\bigwedge_{a,b \in B} [[\![\hat{a} \in U_* \wedge \hat{b} \in U_*]\!] \Rightarrow [\![\hat{a} \wedge_C \hat{b} \in U_*]\!]]$$
$$= \bigwedge_{a,b \in B} [[\![\hat{a} \in U_* \wedge \hat{b} \in U_*]\!] \Rightarrow (a \wedge b)\hat{\ } \in U_*]\!]]$$
$$= \bigwedge_{a,b \in B} [a \wedge b \Rightarrow a \wedge b] = 1.$$

(c) $[\![\forall xy \in C[y \in U_* \wedge y \leq_C x \to x \in U_*]]\!] = 1$. The proof of this is similar to (b), and is left to the reader.

4. Generic Ultrafilters and Models of ZFC   97

(d) $[\![\forall x \in C[x \in U_* \vee x^{\cdot C} \in U_*]]\!]$. The proof of this is also similar to (b).

(e) $[\![U_* \text{ is } (P^{(M)}(B))\hat{\ }\text{-complete}]\!] = 1$. Now the l.h.s. here is

$$[\![\forall X \in (P^{(M)}(B))\hat{\ }[\bigvee X \in U_* \to U_* \cap X \neq \emptyset]]\!]$$
$$= \bigwedge_{A \in P^{(M)}(B)} [\![\bigvee_{C} \hat{A} \in U_* \to U_* \cap \hat{A} \neq \emptyset]\!]$$

and this last expression $= 1$, in view of the fact that, for any $A \in P^{(M)}(B)$,

$$[\![\bigvee_{C} \hat{A} \in U_*]\!] = [\![(\bigvee A)\hat{\ } \in U_*]\!]$$
$$= \bigvee A$$
$$= \bigvee_{a \in A} [\![\hat{a} \in U_*]\!]$$
$$= [\![\exists x \in \hat{A}(x \in U_*)]\!]$$
$$= [\![U_* \cap \hat{A} \neq \emptyset]\!]. \quad \square$$

In view of 4.21, U_* is called the *canonical generic ultrafilter* in $M^{(B)}$ (or in $\hat{B}$): if U is a generic ultrafilter in B, we see from 4.19 that U_* is the natural preimage of U under the map $i_U : M^{(B)} \to M[U]$.

If U is an M-generic ultrafilter in B, $M[U]$ is called a *generic extension* of M. We can now give an *invariant* characterization of $M[U]$ for generic U.

4.22. Theorem. *Let U be an M-generic ultrafilter in B. Then:*

(i) $M[U]$ *is a transitive $\in$-model of* ZFC;

(ii) $M[U]$ *is the least transitive $\in$-model of ZF which includes M and contains U;*

(iii) M *and $M[U]$ have the same ordinals and constructible sets.*

Proof. (i) Since $M[U]$ is, by construction, isomorphic to $M^{(B)}/U$, (i) is an immediate consequence of 4.2.

(ii) Let N be a transitive $\in$-model of ZF such that $M \subseteq N$ and $U \in N$. For each $\alpha \in \text{ORD}^{(M)}$, put $M_\alpha^{(B)} = (V_\alpha^{(B)})^{(M)}$; then clearly

$M_\alpha^{(B)} \in M \subseteq N$ and $M^{(B)} = \bigcup \{M_\alpha^{(B)} : \alpha \in \mathrm{ORD}^{(M)}\}$. Inspection of the definition of the collapsing map i reveals that $i \mid M_\alpha^{(B)}$, the restriction of i to $M_\alpha^{(B)}$, can be defined in N from U; it follows that $\mathrm{ran}(i \mid M_\alpha^{(B)}) \in N$, and, since N is transitive, that $\mathrm{ran}(i \mid M_\alpha^{(B)}) \subseteq N$. Hence
$$M[U] = \bigcup \{\mathrm{ran}(i \mid M_\alpha^{(B)}) : \alpha \in \mathrm{ORD}^{(M)}\} \subseteq N.$$
This proves (ii).

(iii) Let $y \in M[U]$; then $y = i(x)$ for some $x \in M^{(B)}$ and we have

$M[U] \models \mathrm{Ord}[y]$
$\leftrightarrow M[U] \models \mathrm{Ord}[i(x)]$
$\leftrightarrow [\![\mathrm{Ord}(x)]\!] \in U$ (by 4.11)
$\leftrightarrow \bigvee_{\alpha \in \mathrm{ORD}^{(M)}} [\![x = \hat{\alpha}]\!] \in U$ (by (4.5))
$\leftrightarrow \exists \alpha \in \mathrm{ORD}^{(M)} [[\![x = \hat{\alpha}]\!] \in U]$ (since U is generic)
$\leftrightarrow \exists \alpha \in \mathrm{ORD}^{(M)} [y = i(x) = i(\hat{\alpha}) = \alpha]$ (by 4.15)
$\leftrightarrow y \in \mathrm{ORD}^{(M)}.$

Thus M and $M[U]$ have the same ordinals. That M and $M[U]$ have the same constructible sets is proved similarly, using now 1.46 (in M). □

In view of (ii) of this theorem, $M[U]$ may be termed the model of ZFC obtained by *adjoining* U to M, or *generated* by U and M.

Under what conditions do M-generic ultrafilters exist? The following simple example shows that if M is *uncountable* one cannot in general establish the existence of generic ultrafilters. Let M be an uncountable transitive $\in$-model of ZFC such that $\omega_1 \in M$, and put $B = \mathrm{RO}(\omega_1^\omega)^{(M)}$. By Corollary 5.2 to be proved in Chapter 5, $M^{(B)} \models \hat{\omega}_1$ *is countable*. So if U were an M-generic ultrafilter in B, we would have $M[U] \models \omega_1$ *is countable* and hence, since $\omega_1 \in M \subseteq M[U]$, ω_1 would also be countable in the real world. This contradiction shows that there are no M-generic ultrafilters in B.

The situation is quite different, however, when M is *countable*.

4.23. Theorem. *If M is countable, (or, more generally, if $P^{(M)}(B)$ is countable) then for each $b \neq 0$ in B there is an M-generic ultrafilter in B containing b.*

4. Generic Ultrafilters and Models of ZFC

Proof. If M is countable, then so is $P^{(M)}(B)$, and the Rasiowa-Sikorski theorem (Chapter 0) applies. □

4.24. Corollary. *Let σ be any sentence and suppose that in $\mathcal{L}$ we can define a constant term t such that*

$$\text{ZF} + V = L \vdash [t \text{ is a complete Boolean algebra and } V^{(t)} \models \sigma].$$

Then, given any countable transitive $\in$-model of ZF, *we can construct a countable transitive $\in$-model of* ZFC $+ \sigma$.

Proof. Let N be a countable transitive model of ZF, and let M be the submodel of N consisting of all members of N which are constructible in N, i.e. $M = \{x \in N : N \models L[x]\}$. It is well-known that M is then a transitive model of ZF $+ V = L$. Let $B = t^{(M)}$; then B is a complete Boolean algebra in the sense of M and $M^{(B)} \models \sigma$. Let U be an M-generic ultrafilter in B, which exists by 4.23. Then $[\![\sigma]\!]^B = 1 \in U$, so that $M[U] \models \sigma$, and $M[U]$ is the required model. □

It follows immediately from this corollary and the results of Chapter 2 that, given a countable transitive $\in$-model M of ZF, we can construct countable transitive $\in$-models of, e.g.

ZFC + GCH + $P\omega \not\subseteq L$	(2.8, 2.6),
ZFC + $2^{\aleph_0} = \aleph_2 + \forall \kappa \geq \aleph_1 [2^\kappa = \kappa^+]$	(2.19),
ZFC + $2^{\aleph_0} = \aleph_1 + 2^{\aleph_1} = \aleph_{\omega+1}$	(2.20),
ZFC + GCH + $P\omega \subseteq L + P\omega_1 \not\subseteq L$	(2.21).

We conclude this chapter by showing how results about $M^{(B)}$ can be 'transferred' to $V^{(B)}$. The possibility of doing this is based on the following

4.25. Lemma. *Let* Trans(M) *be the formula expressing 'M is transitive'. Let $\phi(x)$ be a formula with one free variable x and suppose that there is a finite conjunction $\tau_1 \wedge \ldots \wedge \tau_n = \tau$ of axioms of ZFC such that*

$$\vdash [\text{Trans}(M) \wedge |M| = \aleph_0 \wedge \tau^{(M)}] \qquad (1)$$
$$\to [\forall B[B \text{ is a complete Boolean algebra} \to \phi(B)]]^{(M)}.$$

Then

$$\text{ZFC} \vdash \forall B[B \text{ is a complete Boolean algebra} \to \phi(B)].$$

Proof. Let σ be the sentence

$$\forall B[B \text{ is a complete Boolean algebra} \to \phi(B)].$$

Then, by the reflection principle and the downward Löwenheim-Skolem theorem, we have

$$\text{ZFC} \vdash \exists M[\text{Trans}(M) \wedge |M| = \aleph_0 \wedge \tau^{(M)} \wedge (\sigma^{(M)} \leftrightarrow \sigma)].$$

The result now follows immediately from (1). □

This lemma enables us to 'transfer' to V results about $V^{(B)}$ derived in M (*i.e.* results about $M^{(B)}$) in the following way. Suppose that $\Phi(V^{(B)})$ is some first-order statement about $V^{(B)}$ which is expressible as a formula $\phi(B)$. Then, if we can show that $\Phi(V^{(B)})$ holds in each countable transitive model of some finite conjunction of axioms of ZFC, it will follow from the lemma that $\Phi(V^{(B)})$ is a theorem of ZFC, *i.e.* $\Phi(V^{(B)})$ 'holds in V'.

The above technique can be applied to elucidate the nature of the canonical generic ultrafilter U_*. First, we note that the same definition of U_* with M replaced by V makes U_* a member of $V^{(B)}$; so we may refer to U_* also as the *canonical generic ultrafilter* in $V^{(B)}$.

Next, we introduce a new constant symbol $\hat{V}$ into $\mathcal{L}^{(B)}$, it being understood that $\hat{V}$ represents a *class*. We extend the assignment of Boolean values to sentences of this augmented language by putting, for $x \in V^{(B)}$,

$$[\![x \in \hat{V}]\!] = \bigvee_{y \in V} [\![x = \hat{y}]\!].$$

Thus $\hat{V}$ represents the class of all *standard* objects in $V^{(B)}$. One can now show that

$$V^{(B)} \models \hat{V} \text{ is a transitive model of ZFC containing all the ordinals}.$$

4. Generic Ultrafilters and Models of ZFC

Moreover

$$V^{(B)} \models (\hat{B} \text{ is a complete Boolean algebra})^{(\hat{V})}.$$

So, working inside $V^{(B)}$, we can construct the $\hat{B}$-extension $\hat{V}^{(\hat{B})}$ of $\hat{V}$. Upon interpreting 4.21 in $V^{(B)}$, with $\hat{V}$ playing the role of M, we see that

$$V^{(B)} \models U_* \text{ is a } \hat{V} - \text{ generic ultrafilter in } \hat{B}.$$

Accordingly, in $V^{(B)}$ we can form the quotient $\hat{V}^{(\hat{B})}/U_*$ and its transitive collapse $\hat{V}[U_*]$.

Applying 4.22 within $V^{(B)}$ (with $\hat{V}$ playing the role of M and $\hat{B}$ that of B) we have

$$V^{(B)} \models \hat{V}[U_*] \text{ is the model of ZFC generated by } U_* \text{ and } \hat{V}.$$

Also, using 4.22 and 4.26 one can show that, for any countable transitive $\in$-model M of ZFC, the statement

$$V^{(B)} \models \forall x(x \in \hat{V}[U_*]) \tag{$*$}$$

holds in M. Moreover, in order to derive $(*)$ in M, one only requires the conjunction of a finite number of axioms of ZFC to hold there. Hence, by the remarks following 4.25, $(*)$ *holds in the real world* (*i.e.* V) for every complete Boolean algebra B. Therefore, if we identify $\hat{V}$ with V, we may regard $V^{(B)}$ as the Boolean-valued model of ZFC generated by U_* and V, or as the Boolean-valued model obtained by *adjoining* the B-valued set U_* to V. It is precisely for this reason that we call $V^{(B)}$ a *Boolean extension* of V.

Problems

Throughout, M is a transitive $\in$-model of ZFC and B is a complete Boolean algebra in the sense of M.

4.26. (*Truth in $M^{(B)}$*). Suppose that M is countable, let $\phi(v_1, \ldots, v_n)$ be a formula and let $x_1, \ldots, x_n \in M^{(B)}$. Show that $M^{(B)} \models \phi(x_1, \ldots, x_n)$ iff

$M[U] \models \phi[i_U(x_1), \ldots, i_U(x_n)]$ for *every* M-generic ultrafilter U, where i_U is the canonical map of $M^{(B)}$ onto $M[U]$. (Use 4.23 and 4.11.) Hence obtain a new proof of 4.21. (Use 4.25.)

4.27. (*Countably M-complete ultrafilters*). An ultrafilter U in B is said to be *countably M-complete* if whenever $X \in P^{(M)}(B)$ is *countable* in M, we have
$$\bigvee X \in U \leftrightarrow X \cap U \neq \emptyset.$$
Put $N = M^{(B)}/U$. Show that the following are equivalent:

(i) U is countably M-complete;

(ii). $\omega^{(N)}$ is well-ordered under $\in_U$;

(ii) $\omega^{(N)} = \{\hat{n}^U : n \in \omega\}$.

(For (i) → (iii), argue as in 4.6. For (ii) → (i), assume (i) fails, choose a partition of unity $\{a_m : m \in \omega\} \in M$ with $a_m \notin U$ for all $m \in \omega$; using the Mixing Lemma define $s_n = \Sigma_{m > n} a_m \cdot (m - n)\hat{\ }$ for each $n \in \omega$. Now show that $\{s_n^U : n \in \omega\}$ is a descending sequence of members of $\omega^{(N)}$.)

4.28. (*Atoms in B*). Let a be an *atom* in B, i.e. such that $a \neq 0$ and, for any $x \in B$, $x \leq a \rightarrow x = 0$ or $x = a$.

(i)   Show that the set $U_a = \{x \in B : a \leq x\}$ is an ultrafilter in B (called the ultrafilter *generated* by a).

(ii) Show that U_a is M-generic, and that $U_a \in M$. Deduce that $M[U_a] = M$.

(iii) Let U be an ultrafilter in B such that $U \in M$, and put $a = \bigwedge U$. Show that the following are equivalent: (a) $a \neq 0$, (b) a is an atom, (c) $U = U_a$.

4.29. (*Atoms and $M[U]$*). Let A be the set of all atoms in B, and let U_* be the canonical generic ultrafilter in $M^{(B)}$.

(i)   Show that $\bigvee_{y \in M} [\![U_* = \hat{y}]\!] = \bigvee A$. (Observe that $[\![U_* = \hat{y}]\!] \leq [\![\hat{y}$ is an ultrafilter in $\hat{B}]\!] = 0$ or 1, and use 4.28(iii).)

(ii) Show that $\bigwedge_{x \in M^{(B)}} \bigvee_{y \in M} [\![x = \hat{y}]\!] = \bigvee A$. (For any atom $a \in B$, and any $x \in M^{(B)}$, show, using 4.28(i) and 4.11 that $a \leq \bigvee_{y \in M} [\![x = \hat{y}]\!]$.)

(iii) Assume that $M \models V = L$. Show that

$$[\![V = L]\!] = \bigwedge_{x \in M^{(B)}} \bigvee_{y \in M} [\![x = \hat{y}]\!] = \bigvee_{y \in M} [\![U_* = \hat{y}]\!] = [\![L(U_*)]\!].$$

(Use 1.46 and (i).)

(iv) Put $\eta = 1$ if $M \models V = L$ and $\eta = 0$ if $M \models V \neq L$. Show that $[\![V = L]\!] = \eta \wedge \bigvee A$.

(v) Assume that $M \models V = L$, and let U be an M-generic ultrafilter in B. Show that $M[U] \models V = L$ iff $U = U_a$ for some atom $a \in B$. (Use (iv).)

4.30. (*A trivial Boolean extension*). Let $u \in M$, and put $B = P^{(M)}(u)$. B is then the power set Boolean algebra of u in M.

(i) Show that, for any formula $\phi(v_1, \ldots, v_n)$, and any $x_1, \ldots, x_n \in M$,

$$M \models \phi[x_1, \ldots, x_n] \leftrightarrow M^{(B)} \models \phi(\hat{x}_1, \ldots, \hat{x}_n).$$

(If $[\![\phi(\hat{x}_1, \ldots, \hat{x}_n)]\!] \neq 1$, let a be an atom $\leq [\![\neg \phi(\hat{x}_1, \ldots, \hat{x}_n)]\!]$, and use 4.28(ii).)

(ii) Let U be any ultrafilter in B. Show that, for any sentence σ,

$$M \models \sigma \leftrightarrow M^{(B)}/U \models \sigma.$$

4.31. (*A transitive model of* $\neg AC$). Let $G \in M$ be a group acting on B, and let $\Gamma \in M$ be a filter of subgroups of G. Put $M^{(\Gamma)} = (V^{(\Gamma)})^{(M)}$ (for the definition of $V^{(\Gamma)}$, see Chapter 3). Let U be an M-generic ultrafilter in B. Recalling that i is the natural map of $M^{(B)}$ onto $M[U]$, put $M[\Gamma, U] = \langle i[M^{(\Gamma)}], \in | \, i[M^{(\Gamma)}] \rangle$.

(i) Show that $M \subseteq M[\Gamma, U]$, and that $M[\Gamma, U]$ is transitive.

(ii) Show that, for any formula $\phi(v_1, \ldots, v_n)$, and any $x_1, \ldots, x_n \in M^{(\Gamma)}$,

$$M[\Gamma, U] \models \phi[i(x_1), \ldots, i(x_n)] \leftrightarrow [\![\phi(x_1, \ldots, x_n)]\!]^\Gamma \in U.$$

(iii) Show that, if M is countable, then for a suitable choice of B, Γ and U, $M[\Gamma, U]$ is a countable transitive model of ZF in which AC fails. (Use 3.21)

4.32.[†] (*The converse to 4.7 fails*). Suppose that there is a measurable cardinal $\mu > \omega$ and an inaccessible cardinal $\kappa > \mu$. Then it follows that $M = \langle R_\kappa, \in | R_\kappa \rangle$ is a transitive $\in$-model of ZFC. Let U be a μ-complete non-principal ultrafilter in $P\mu \in M$. Show that $M^{(P\mu)}/U$ is well-founded but U is not M-generic. (Note that, by 3.13, $M^{(P\mu)}/U$ is isomorphic to the ultrapower M^μ/U.)

4.33. (*Construction of uncountable transitive models of* ZFC $+ V \neq L$).

(i) Let κ be an infinite cardinal, and suppose that the complete Boolean algebra B contains a κ-closed dense subset P (2.17). Show that, for any $S \subseteq PB$ such that $|S| \leq \kappa$, there is an S-complete ultrafilter in B. (Let $S = \{T_\xi : \xi < \kappa\}$ and first confine attention to the case in which $\bigvee T_\xi = 1$ for all $\xi < \kappa$. Let J be a sufficiently large index set so that each T_α can be enumerated as $\{t_{\xi j} : j \in J\}$. Using the fact that P is κ-closed, construct by transfinite recursion a function $f \in J^\kappa$ such that $\bigwedge_{\xi < \alpha} t_{\xi f(\xi)} \neq 0$ for each $\alpha < \kappa$. Conclude that there is an ultrafilter which intersects each T_ξ; now apply this to the general case.)

(ii) Let κ be a regular cardinal. Show that, for each family S of subsets of $B_\kappa(\kappa, 2)$ (2.18) such that $|S| \leq \kappa$, there is an S-complete ultrafilter in $B_\kappa(\kappa, 2)$. (Use (i) and 2.18(ii).)

(iii) Suppose that there exists an inaccessible cardinal $\lambda > \omega$. Show that, for each infinite cardinal $\kappa < \lambda$ there is a transitive $\in$-model of ZFC $+ V \neq L$ of cardinality κ. (By Löwenheim-Skolem it is enough to prove the result for all *regular* $\kappa < \lambda$. So let κ be regular, put

---

[†]This problem assumes an acquaintance with measurable cardinals; cf. Drake (1974).

4. Generic Ultrafilters and Models of ZFC

$B = B_\kappa(\kappa, 2)$ and $M = \langle R_\lambda, \in | R_\lambda \rangle$. Using the Maximum Principle in $M^{(B)}$, for each formula $\phi(v_0, \ldots, v_n)$ let $f_\phi : (M^{(B)})^n \to M^{(B)}$ be a 'Skolem function for ϕ in $M^{(B)}$', i.e., such that, for all $x_1, \ldots, x_n \in M^{(B)}$, $[\![\exists v_0 \phi(v_0, x_1, \ldots, x_n)]\!] = [\![\phi(f_\phi(x_1, \ldots, x_n), x_1, \ldots, x_n)]\!]$. Let Λ be the closure of the set $\{\hat{\xi} : \xi < \kappa\}$ under all the f_ϕ. By (ii), let U be an S-complete ultrafilter in B, where $S = \{\{[\![a = \hat{\xi}]\!] : \xi < \kappa\} : a \in \Lambda\}$. Show that the structure $\langle \{a^U : a \in \Lambda\}, \in_U \rangle$ is a well-founded model of ZFC $+ V \neq L$ of cardinality κ.)

4.34. (*Generic sets of conditions*). Let $\langle P, \leq \rangle$ be a partially ordered set in M. A subset X of P is said to be *dense* in P if $\forall y \in P \exists x \in X [x \leq y]$, and *dense below an element* $p \in P$ if $\forall y \leq p \exists x \in X [x \leq y]$. A subset G of P is said to be *M-generic* if

(a) $x \in G, y \in P, x \leq y \to y \in G$;

(b) $\forall xy \in G \exists z \in G [z \leq x \wedge z \leq y]$;

(c) $G \cap X \neq \emptyset$ for every dense subset X of P which is in M.

(i) Show that, if G is M-generic in P and $X \subseteq P$ is dense below an element of G, then $X \cap G \neq \emptyset$.

Now let Q be the refined associate of P (2.4) and let $j : P \to Q$ be the canonical map (if P is refined then $P = Q$ and j is the identity). Let $B = \mathrm{RO}(Q)^{(M)}$ be the Boolean completion of Q in M, and identify Q as a dense subset of B.

(ii) Show that, if G is M-generic in P, then

$$\overline{G} = \{x \in B : \exists y \in G [j(y) \leq x]\}$$

is an M-generic ultrafilter (called the M-generic ultrafilter *generated by G*). Show also that $G = j^{-1}[\overline{G}]$. Show conversely, that if U is an M-generic ultrafilter in B, then $j^{-1}[U]$ is an M-generic subset of P. Deduce that, if M is *countable*, P has an M-generic subset.

It follows from (ii) that, if U is the M-generic ultrafilter generated by an M-generic subset G of P, then G is definable from U and vice-versa. *Under these circumstances we write $M[G]$ for $M[U]$.* If $p \in P$, and σ is an

$\mathcal{L}_M^{(B)}$-sentence, we write $p \Vdash \sigma$ for $j(p) \Vdash \sigma$.

(iii) Show that, if G is M-generic in P, $M[G]$ is the least transitive model of ZF which includes M and contains G.

(iv) Let G be M-generic in P and let i be the canonical map of $M^{(B)}$ onto $M[G]$. Show that, for any formula $\phi(v_1, \ldots, v_n)$ and any $x_1, \ldots, x_n \in M^{(B)}$,

$$M[G] \models \phi[i(x_1), \ldots, i(x_n)] \leftrightarrow \exists p \in G[p \Vdash \phi(x_1, \ldots, x_n)].$$

(v) *Iteration lemma.* Let P be a refined partially ordered set in M, let G be an M-generic subset of P, let Q be a refined partially ordered set in $M[G]$, and let H be an $M[G]$-generic subset of Q. Show that there is a refined partially ordered set R in M and an M-generic subset K of R such that $M[G][H] = M[K]$. (Let $\leq_Q$ be the partial ordering of Q. First show that without loss of generality we may assume that Q (but not $\leq_Q$) is in M. Let B be the Boolean completion of P in M, let i be the canonical map of $M^{(B)}$ onto $M[G]$, let $\leq^*$ be an element of $M^{(B)}$ such that $i(\leq^*) = \leq_Q$, and let σ be the $\mathcal{L}_M^{(B)}$-sentence: $\langle \hat{Q}, \leq^* \rangle$ *is a refined partially ordered set.* Observe that there is $p \in G$ for which $p \Vdash \sigma$. Now let $P' = \{p \in P : p \Vdash \sigma\}$, $R = P' \times Q$ and define the relation $\leq$ on R by $\langle p_1, q_1 \rangle \leq \langle p_2, q_2 \rangle$ iff ($p_1 \leq p_2$ and $p_1 \Vdash \hat{q}_1 \leq^* \hat{q}_2$). Show that $\langle R, \leq \rangle$ is a refined partially ordered set in M. Now put $G' = G \cap P'$ and $K = G' \times H$. Show that every dense subset of P' meets G'; use this, together with the genericity of G and H to prove that K is M-generic. Finally, use the fact that $G = \{p \in P : \exists r \in G'[r \leq p]\}$ to prove the last assertion.)

(vi) *Product lemma.* Let P and Q be refined partially ordered sets in M, let G be an M-generic subset of P and let H be an $M[G]$-generic subset of Q. Give $P \times Q$ the product ordering: $\langle p_1, q_1 \rangle \leq \langle p_2, q_2 \rangle$ iff $p_1 \leq p_2$ and $q_1 \leq q_2$. Show that $G \times H$ is an M-generic subset of $P \times Q$ and that $M[G \times H] = M[G][H]$. (Like (v).)

4.35. (*Canonical generic sets and the adjunction of maps*). Let P be a basis for B in M. Define $G_* \in M^{(B)}$ by $\text{dom}(G_*) = \{\hat{p} : p \in P\}$, $G_*(\hat{p}) = p$ for $p \in P$.

4. Generic Ultrafilters and Models of ZFC

(i) Let G be M-generic in P, and let $i: M^{(B)} \to M[G]$ be the canonical map. Show that $i(G_*) = G$. (Like 4.19.)

(ii) Show that
$$M^{(B)} \models G_* \text{ is a generic subset of } \hat{P}.$$

(Like 4.21.) For this reason G_* is called the *canonical generic set* in $M^{(B)}$.

Now define $G_{**} \in M^{(B)}$ by $\mathrm{dom}(G_{**}) = \bigcup \{\mathrm{dom}(y) : y \in \mathrm{dom}(G_*)\}$ and $G_{**}(x) = [\![\exists y \in G_*[x \in y]]\!]$ for $x \in \mathrm{dom}(G_{**})$.

(iii) Show that $M^{(B)} \models G_{**} = \bigcup G_*$.

Now suppose that a, b are non-empty elements of M such that $|b| \geq 2$ and $\aleph_0 \leq |a|$ in M. Let G be an M-generic subset of $P = C(a,b)^{(M)}$ and let $B = \mathrm{RO}(b^a)^{(M)}$.

(iv) Show that $M^{(B)} \models G_{**}$ *is a map of* $\hat{a}$ *onto* $\hat{b}$. (See the proof of 5.1.)

(v) Show that $M[G] \models \bigcup G$ *is a map of* a *onto* b.

Results (iv) and (v) show that, for this choice of B and G, in $M^{(B)}$ we have adjoined the canonical 'map' G_{**} of $\hat{a}$ onto $\hat{b}$ and in $M[G]$ we have adjoined the map $\bigcup G$ of a onto b. Notice that if a is (really) *countable* and b is (really) *uncountable*, no transitive model of ZF can contain a map of a onto b. It follows that, if M is uncountable, there may be no M-generic subset of $C(a,b)^{(M)}$ and hence no M-generic ultrafilters in $\mathrm{RO}(b^a)^{(M)}$. On the other hand,

(vi) If M is *countable*, show that there is an M-generic subset of P. (Use 4.23 and 4.34(ii).)

(vii) Let M be countable, put $P = C(\omega, 2)^{(M)}$ and let G be an M-generic subset of P. Show that, in $M[G]$, $\bigcup G$ is a non-constructible map of ω into 2. (Like 2.6.)

(viii) Let M be countable, and suppose that $M \models \mathrm{GCH}$. Put $P = C(\omega \times \omega_2, 2)^{(M)}$ and let G be an M-generic subset of P. Show that, in $M[G]$, $\bigcup G$ is a map of $\omega \times \omega_2^{(M)}$ onto 2. For each $\nu < \omega_2^{(M)}$, put $u_\nu = \{n \in \omega : (\bigcup G)(n, \omega) = 1\}$. Show that $\{u_\nu : \nu < \omega_2^{(M)}\}$ is a set of $\aleph_2^{(M)} = \aleph_2^{(M[G])}$ subsets of ω in $M[G]$. (Like 2.12.)

4.36. *(Adjunction of a subset of ω).* Let P be a basis for B in M, and let G be an M-generic subset of P. If $s \in M[G]$, $s \subseteq m$ (or equivalently, if $s \in M[U]$, $s \subseteq M$ where U is the M-generic ultrafilter generated by G, cf. 4.34), it can be shown—although we do not prove it here (for a proof see Grigorieff 1975)—that there is a least transitive model $M[s]$ of ZF which includes $M \cup \{s\}$. $M[s]$ is called the model of ZF obtained by *adjoining s to M*: it has the following important properties:

(1) $M[s] \subseteq M[G]$;

(2) $M[s] \models AC$;

(3)   if $t \in M[G]$ is absolutely definable from s and elements of M, then $t \in M[s]$.

(i)   Let $a, b \in M$; let G be an M-generic subset of $C(a,b)^{(M)}$ and let $F = \bigcup G$. Show that $M[F] = M[G]$. (Note that we have $G = \{f \in C(a,b)^{(M)} : f \subseteq F\}$.)

(ii)   Let κ be an infinite cardinal in M; let G be a generic subset of $C(\omega, \kappa)^{(M)}$, and let $F = \bigcup G$. Show that there is $s \subseteq \omega$, $s \in M[F]$ such that $M[F] = M[s]$. (Put $s = \{2^n 3^m : F(n) \leq F(m)\} \in M[F]$. Put $m \sim n$ iff $F(m) = F(n)$; the set A of equivalence classes $\tilde{m}$ can be ordered by $\tilde{m} < \tilde{n}$ iff $2^n 3^m \notin s$. Show that F induces an order-preserving map $F' : \langle A, < \rangle \to \langle \kappa, < \rangle$, so that $\langle A, < \rangle$ is well-ordered in $M[s]$. Let $F'' : A \to \alpha$ be an isomorphism of A with an ordinal in $M[s]$. Show that $F'' \circ (F')^{-1}$ is the identity, and conclude that $F \in M[s]$.)

4.37. *(Intermediate submodels and complete subalgebras).* Let $X \in P^{(M)}(B)$. The complete subalgebra (in M) of B *generated* by X is defined to be the least complete subalgebra of B in M which includes X.

(i)   Suppose that $|B| = \kappa$ (in M), and let B' be the complete subalgebra of B generated by X. Define the sets $\{B_\alpha : \alpha < \kappa^{+(M)}\}$ induc-

4. Generic Ultrafilters and Models of ZFC

tively as follows: $B_0 = X$; if α is odd, $B_\alpha = \{b^* : b \in \bigcup_{\beta<\alpha} B_\beta\}$; if α is even, $B_\alpha = \{\bigvee X : X \in M \text{ and } X \subseteq \bigcup_{\beta<\alpha} B_\beta\}$; if α is a limit, $B_\alpha = \bigcup_{\beta<\alpha} B_\beta$. Show that $B' = \bigcup_{\beta<\alpha} B_\beta$.

From now on we let U be an M-generic ultrafilter in B, and let $i : M^{(B)} \to M[U]$ be the canonical map.

(ii) Let $s \in M[U]$ and $s \subseteq M$. Show that there is $t \in M$ such that $s \subseteq t$, and hence $s_* \in M^{(B)}$ such that $i(s_*) = s$ and $\text{dom}(s_*) = t$. (For each $x \in M[U]$ let $\underline{x} \in M^{(B)}$ be such that $i(\underline{x}) = x$. Then $s \subseteq M$ means that $\bigvee_{y \in M} [\![\underline{x} = \hat{y}]\!] \in U$ for each $x \in s$. Now argue as in the proof of 1.36. For the second assertion, consult the proof of 1.38.)

(iii) Let $s \in M[U]$ and $s \subseteq M$. Let $B(s_*)$ be the complete subalgebra of B generated by $\{[\![\hat{x} \in s_*]\!] : x \in M\}$, where s_* is as in (ii). Show that $M[s] = M[U \cap B(s_*)]$. (Notice first that the equation in question makes sense because $U \cap B(s_*)$ is an M-generic ultrafilter in $B(s_*)$ and $s_* \in M^{(B(s_*))}$. Next, show that $s \in M[U \cap B(s_*)]$, so that $M[s] \subseteq M[U \cap B(s_*)]$. Now prove the reverse inclusion by showing that, if N is any transitive model of ZF and $M \cup \{s\} \subseteq N$, then $U \cap B(s_*) \in N$. Put $U_\alpha = U \cap B_\alpha$, where the B_α are defined as in (i), with $X = \{[\![\hat{x} \in s_*]\!] : x \in M\}$. Show by induction that $U_\alpha \in N$ for all $\alpha < (|B|^+)^{(M)}$. Conclude that $U \cap B(s_*) = \bigcup_\alpha B_\alpha \in N$.)

(iv) Let N be a transitive model of ZFC such that $M \subseteq N \subseteq M[U]$. Show that $N = \bigcup \{M[s] : s \subseteq M \text{ and } s \in N\}$. (Given $x \in N$, show that there is $s \in N$ with $s \subseteq \text{ORD}^{(N)} = \text{ORD}^{(M)}$ such that $x \in M[s]$ as follows. Since AC holds in N, there is a bijection f in N of a cardinal $\kappa \in N$ onto the transitive closure t of x. Define $r \subseteq \kappa \times \kappa$ by $\langle \alpha, \beta \rangle \in r \leftrightarrow f(\alpha) \in f(\beta)$. Let g be the canonical map of $\kappa \times \kappa$ onto κ (in M), and put $s = g[r]$.)

(v) Let N be a transitive model of ZFC such that $M \subseteq N \subseteq M[U]$. Show that there is a complete subalgebra A of B such that $N = M[U \cap A]$. (By (iv), choose $s \subseteq M$, $s \in N$ such that $P^{(N)}(B) \in M[s] \subseteq N$. Now use (iii) and (iv) to get $N = M[s]$ and apply (iii) again.)

(vi) B is said to be *countably generated* (in M) if there is a countable subset X of B in M such that the complete subalgebra of B generated by X is B itself. Show that, if B is countably generated, then there is an $s \in M[U]$, $s \subseteq \omega$ such that $M[U] = M[s]$. (Let $X = \{b_n : n \in \omega\}$ be the countable generating set. Define $s' \in M^{(B)}$ by $\text{dom}(s') = \hat{\omega}$ and $s'(\hat{n}) = b_n$. Use (iii) to show that $s = i(s')$ meets the requirements.)

4.38. (*Involutions and generic ultrafilters*). Let U be an M-generic ultrafilter in B and let $\pi \in M$ be an automorphism of B.

(i) Show that $\pi[U]$ is an M-generic ultrafilter in B and that $M[U] = M[\pi[U]]$.

(ii) Let $f : B \to B$, $f \in M$ be such that $f[U] \subseteq U$. Show that there is a $b \in U$ such that $f(x) \geq x$ for all $x \leq b$. (Put $b = \bigwedge \{x^* \vee f(x) : x \in B\}$.)

(iii) An *M-involution* of B is an automorphism $\pi \in M$ of B such that π^2 is the identity. Let $W \subseteq B$, $W \in M[U]$. Show that the following conditions are equivalent:

(a) W is an M-generic ultrafilter in B and $M[U] = M[W]$;

(b) there is an M-involution π of B such that $\pi[U] = W$. (For (b) $\to$ (a), use (i). For (a) $\to$ (b), assume (a) and $U \neq W$. Let i_U, i_W be the canonical maps of $M^{(B)}$ onto $M[U]$, $M[W]$ respectively and let $\underline{U}, \underline{W}$ be such that $i_W(\underline{U}) = U$, $i_U(\underline{W}) = W$. Define the functions k, l, m, n of B into B by $k(x) = [\![\hat{x} \in \underline{U}]\!]$, $m(x) = [\![\hat{x} \in \underline{W}]\!]$, $l(x) = \bigwedge \{y \in B : x \leq m(y)\}$, $n(x) = \bigwedge \{y \in B : x \leq k(y)\}$. Now put $f(x) = k(x) \wedge l(x)$, $g(x) = m(x) \wedge n(x)$. Show that $(g \circ f)[U] \subseteq U$ and $(f \circ g)[W] \subseteq W$ and that $(g \circ f)(x) \leq x$, $(f \circ g)(x) \leq x$. By (ii), choose $b_0 \in U$, $c_0 \in W$ such that $(g \circ f)(x) = x$ for all $x \leq b_0$ and $(f \circ g)(x) = x$ for all $x \leq c_0$. Since $U \neq W$, we may assume that $b_0 \wedge c_0 = 0$. Put $b = b_0 \wedge g(c_0) \in U$ and $c = f(b) \in W$, and let $B_b = \{x \in B : x \leq b\}$, $B_c = \{x \in B : x \leq c\}$. Show that $f \mid B_b$ is an isomorphism of B_b onto B_c and $g \mid B_c$ is the inverse of $f \mid B_b$. Now put, for $x \in B$, $\pi(x) = f(x \wedge b) \vee g(x \wedge c) \vee (x - (b \vee c))$.)

4.39. (*The submodel of hereditarily ordinal definable sets*). Let $\mathcal{L}_S$ be the extension of the language $\mathcal{L}$ of set theory to include a new unary predicate symbol S, and let $\mathcal{L}_S^{(B)}$ be the language obtained from $\mathcal{L}_M^{(B)}$ by adding S. We extend the assignment of Boolean values to $\mathcal{L}_S^{(B)}$-sentences by defining, for $x \in M^{(B)}$, $[\![S(x)]\!] = \bigvee_{y \in M} [\![x = \hat{y}]\!]$. Let U be an M-generic ultrafilter in B, and let i_U be the canonical map of $M^{(B)}$ onto $M[U]$. We write $(M[U], M)$ for the $\mathcal{L}_S$-structure obtained from $M[U]$ by interpreting S as the subset M of $M[U]$.

  (i) Show that for any $\mathcal{L}_S$-formula $\phi(v_1, \ldots, v_n)$ and any $x_1, \ldots, x_n \in M^{(B)}$

$$(M[U], M) \models \phi[i_U(x_1), \ldots, i_U(x_n)] \leftrightarrow [\![\phi(x_1, \ldots, x_n)]\!] \in U.$$

Now put $(\text{ODM})^{M[U]}$ for the collection of all elements of $M[U]$ which are definable in $(M[U], M)$ from (ordinals and) elements of M, $(\text{HODM})^{M[U]}$ for the collection of all elements of $(\text{ODM})^{M[U]}$ whose transitive closure is in $(\text{ODM})^{M[U]}$. $(\text{ODM})^{M[U]}$ and $(\text{HODM})^{M[U]}$ are the sets of elements of $M[U]$ which are ordinal definable, and hereditarily ordinal definable, respectively, from M in $M[U]$. It can be shown that $(\text{HODM})^{M[U]}$ is a transitive model of ZFC; evidently $M \subseteq (\text{HODM})^{M[U]} \subseteq M[U]$, so by 4.37(v) there is a complete subalgebra A of B such that $(\text{HODM})^{M[U]} = M[U \cap A]$. We now describe A explicitly.

Let $B^+ = \{x \in B : \pi(x) = x \text{ for every automorphism } \pi \in M \text{ of } B\}$. It is easy to see that B^+ is a complete subalgebra of B.

  (ii) Show that $(\text{HODM})^{M[U]} = M[U \cap B^+]$. (Put $X = \{W \in M[U] : W \text{ is an } M\text{-generic ultrafilter in } B \text{ and } M[U] = M[W]\}$. By 4.38 $X = \{\pi[U] : \pi \in M \text{ is an automorphism of } B\}$. Hence show that $U \cap B^+ = \bigcap\{W \cap B^+ : W \in X\}$ and that $U \cap B^+$ is definable in $(M[U], M)$ from B. Infer that $U \cap B^+ \in (\text{HODM})^{M[U]}$, so that $M[U \cap B^+] \subseteq (\text{HODM})^{M[U]}$. To establish the reverse inclusion, observe that it suffices to show that, if $s \in (\text{HODM})^{M[U]}$ and $s \subseteq M[U \cap B^+]$, then $s \in M[U \cap B^+]$ (for if $(\text{HODM})^{M[U]} - M[U \cap B^+] \neq \emptyset$, an element of it of minimal rank would be a subset of $M[U \cap B^+]$). Since $U \cap B^+$ is definable in $(M[U], M)$ from B, so is the canonical map of $M^{(B^+)}$ onto $M[U \cap B^+]$, so we may suppose that $s \subseteq M$. Let $t \in M$ be such that $s \subseteq t$, let $x_0, \ldots, x_n \in M$ and let $\phi(v_0, \ldots, v_{n+1})$ be an $\mathcal{L}_S$-formula such that for all $x \in t$,

$$x \in s \leftrightarrow (M[U], M) \models \phi(x, x_0, \ldots, x_n).$$

Notice that $[\![\phi(\hat{x}, \hat{x}_0, \ldots, \hat{x}_n)]\!] \in B^+$, and that $x \in s$ if and only if $[\![\phi(\hat{x}, \hat{x}_0, \ldots, \hat{x}_n)]\!] \in U \cap B^+$. Conclude that $s \in M[U \cap B^+]$.)

Chapter 5

Cardinal Collapsing and Some Applications to the Theory of Boolean Algebras

Cardinal Collapsing

We have seen in Chapter 1 that if the complete Boolean algebra B satisfies the countable chain condition, then cardinals in V retain their true size in $V^{(B)}$. In this section we show that, if B does not satisfy this condition, it is sometimes possible to have two infinite cardinals $\kappa < \lambda$ for which $V^{(B)} \models |\hat{\lambda}| = |\hat{\kappa}|$. In this event we say that λ has been *collapsed* to κ in $V^{(B)}$. We begin by formulating a necessary and sufficient condition on B for this to happen.

5.1. Theorem. *Let κ and λ be infinite cardinals with $\kappa \leq \lambda$. Then the following conditions are equivalent:*

(i) $V^{(B)} \models |\hat{\kappa}| = |\hat{\lambda}|$;

(ii) *there is a double sequence $\{b_{\xi\eta} : \xi < \kappa, \eta < \lambda\} \subseteq B$ such that $\bigvee_{\xi < \kappa} b_{\xi\eta} = 1$ for all $\eta < \lambda$ and $\{b_{\xi\eta} : \eta < \lambda\}$ is an antichain for each $\xi < \kappa$.*

Proof. (i) → (ii). Suppose (i) holds. Then we have

$$V^{(B)} \models \exists f[f \text{ is a map of } \hat{\kappa} \text{ onto } \hat{\lambda}].$$

Using the Maximum Principle, it follows that there is $f \in V^{(B)}$ such that

$$V^{(B)} \models f \text{ is a map of } \hat{\kappa} \text{ onto } \hat{\lambda}. \tag{1}$$

Put $b_{\xi\eta} = [\![f(\hat{\xi}) = \hat{\eta}]\!]$ for $\xi < \kappa, \eta < \lambda$. Then, if $\eta, \eta' < \lambda$ and $\eta \neq \eta'$,

$$b_{\xi\eta} \wedge b_{\xi\eta'} = [\![f(\hat{\xi}) = \hat{\eta} \wedge f(\hat{\xi}) = \hat{\eta}']\!] \leq [\![\hat{\eta} = \hat{\eta}']\!] = 0,$$

and, for $\eta < \lambda$,

$$\bigvee_{\xi < \kappa} b_{\xi\eta} = \bigvee_{\xi < \kappa} [\![f(\hat{\xi}) = \hat{\eta}]\!] = [\![\exists x \in \hat{\kappa}[f(x) = \hat{\eta}]]\!] = 1,$$

by (1). Thus $\{b_{\xi\eta} : \xi < \kappa, \eta < \lambda\}$ satisfies (ii).

(ii) $\to$ (i). Assume (ii). Since $\kappa < \lambda$, we have $V^{(B)} \models |\hat{\kappa}| \leq |\hat{\lambda}|$, so it suffices to show that $V^{(B)} \models |\hat{\lambda}| \leq |\hat{\kappa}|$. To this end, define $f \in V^{(B)}$ by

$$\mathrm{dom}(f) = \{\langle \hat{\xi}, \hat{\eta}\rangle^{(B)} : \xi < \kappa, \eta < \lambda\}$$

and, for $\xi < \kappa, \eta < \lambda$,

$$f(\langle \hat{\xi}, \hat{\eta}\rangle^{(B)}) = b_{\xi\eta}.$$

Using the assumption that $\{b_{\xi\eta} : \eta < \lambda\}$ is an antichain for each $\xi < \kappa$, it follows easily that

$$V^{(B)} \models f \text{ is a map with } \mathrm{dom}(f) \subseteq \hat{\kappa} \text{ and } \mathrm{ran}(f) \subseteq \hat{\lambda}.$$

Also, for each $\eta < \lambda$ we have, by assumption,

$$[\![\exists x \in \hat{\kappa}[f(x) = \hat{\eta}]]\!] = \bigvee_{\xi < \kappa} [\![f(\hat{\xi}) = \hat{\eta}]\!] = \bigvee_{\xi < \kappa} b_{\xi\eta} = 1.$$

It follows that $V^{(B)} \models \hat{\lambda} \subseteq \mathrm{ran}(f)$, and so $V^{(B)} \models |\hat{\lambda}| \leq |\hat{\kappa}|$, completing the proof. □

Let λ be an infinite cardinal, let X be the product space λ^ω, where λ is assigned the discrete topology, and let $B = \mathrm{RO}(X)$. For each $m \in \omega$ and $\eta < \lambda$ let $b_{m\eta} = \{g \in X : g(m) = \eta\}$. It is now straightforward to verify that $\{b_{m\eta} : m \in \omega, \eta < \lambda\}$ is a subset of B which satisfies 5.1(ii), with $\kappa = \omega$. Accordingly, that theorem gives:

5.2. Corollary. *Let $B = \mathrm{RO}(\lambda^\omega)$, where $\lambda \geq \aleph_0$. Then*

$$V^{(B)} \models \hat{\lambda} \text{ is countable }. \quad \square$$

5. Cardinal Collapsing

This result shows that $\text{RO}(\lambda^\omega)$ may be thought of as an algebra which adjoins a collapsing map of $\hat{\omega}$ onto $\hat{\lambda}$; accordingly $\text{RO}(\lambda^\omega)$ is called the *collapsing* $(\aleph_0, \lambda)$-*algebra*. In the next section we shall show that these collapsing algebras have other useful features.

Problems

5.3. $(P\omega \cap L$ *can be countable*).

(i) Let $\lambda \geq \aleph_0$ and let B be the collapsing $(\aleph_0, 2^\lambda)$-algebra. Show that
$$V^{(B)} \models P\hat{\lambda} \cap L \text{ is countable}.$$

(ii) Let M be a countable transitive model of $\text{ZFC} + 2^{\aleph_0} = \aleph_1$, put $B = (\text{RO}(\omega_1^\omega))^{(M)}$, and let U be an M-generic ultrafilter in B. Show that
$$M[U] \models P\omega \cap L \text{ is countable}.$$

5.4. (*More on collapsing algebras*). Assume GCH. Let κ, λ be regular infinite cardinals with $\kappa < \lambda$. Put $B = B_\kappa(\kappa, \lambda)$ (cf. 2.18).

(i) Show that $V^{(B)} \models \text{Card}(\hat{\alpha})$ for any cardinal $\alpha \leq \kappa$. (Use 2.20(i) and 2.18.)

(ii) Show that $V^{(B)} \models \text{Card}(\hat{\alpha})$ for any cardinal $\alpha \geq \lambda^+$. (Show that $|C_\kappa(\kappa, \lambda)| = \lambda$ and so B satisfies the $\lambda^+ - \text{cc}$. Now use 1.53.)

(iii) Show that $V^{(B)} \models |\hat{\lambda}| = \hat{\kappa}$. (Use 5.1.)

$B_\kappa(\kappa, \lambda)$ is called the *collapsing* (κ, λ)-*algebra*: its effect is to collapse λ to κ but not to collapse any cardinal $\leq \kappa$ or $\geq \lambda^+$.

5.5. (*Consistency of* CH *and* $\neg$CH *with the existence of measurable cardinals*). Let κ be a cardinal. An ultrafilter F in $P\kappa$ is said to be *nonprincipal* if $\{\alpha\} \notin F$ for all $\alpha < \kappa$, and κ-*complete* if whenever $\alpha < \kappa$

and $\{X_\xi : \xi < \alpha\} \subseteq F$, then $\bigwedge_{\xi<\alpha} X_\xi \in F$. The cardinal κ is said to be *measurable* (cf. Drake 1974) if $\kappa > \aleph_0$ and there is a non-principal κ-complete ultrafilter in $P\kappa$. It is known that, if κ is measurable, then κ is regular and $2^\lambda < \kappa$ for every cardinal $\lambda < \kappa$, i.e. κ is inaccessible.

(i) Let κ be a measurable cardinal and let F be a κ-complete non-principal ultrafilter in $P\kappa$. Let B be a complete Boolean algebra with a basis P such that $|P| < \kappa$. Define $G \in V^{(B)}$ by $\mathrm{dom}(G) = B^{\mathrm{dom}(\hat\kappa)}$ and, for $y \in \mathrm{dom}(G)$, $G(y) = [\![y \subseteq \hat\kappa \wedge \exists x \in \hat{F}[x \subseteq y]]\!]$.

(a) Show that $V^{(B)} \models \mathrm{Card}(\hat\kappa) \wedge \hat\kappa > \aleph_0$. (Use 1.53(iii).)

(b) Show that $V^{(B)} \models G$ *is a non-principal filter in* $P\hat\kappa$. (Use the fact that $[\![u \in G]\!] = [\![u \subseteq \hat\kappa]\!] \wedge \bigvee_{x \in F} [\![\hat x \subseteq u]\!]$.)

(c) Show that $V^{(B)} \models G$ *is an ultrafilter in* $P\hat\kappa$. (For this it suffices to show that, for any $p \in P$ and $u \in V^{(B)}$, $p \Vdash (u \cap \hat\kappa \in G) \vee (\hat\kappa - u \in G)$. Let $t = \{\alpha < \kappa : p \Vdash \hat\alpha \in u\}$. Show that, if $t \in F$, then $p \Vdash u \cap \hat\kappa \in G$. On the other hand, if $t \notin F$, then $\{\alpha < \kappa : p \not\Vdash \hat\alpha \in u\} \in F$; using the κ-completeness of F and the fact that $|P| < \kappa$, deduce that there is $q \leq p$ such that $q \Vdash \hat\kappa - u \in G$. Now use 2.5(iii).)

(d) Show that $V^{(B)} \models G$ *is* $\hat\kappa$-*complete*. (Given $\alpha < \kappa, p \in P$ and $p \Vdash f : \hat\alpha \to G$, it must be shown that $p \Vdash \bigcap_{x \in \hat\alpha} f(x) \in G$. For each $\xi < \alpha$, put $t_\xi = \{\beta < \kappa : p \Vdash \hat\beta \in f(\hat\xi)\}$. Argue as in (c) to derive a contradiction from the assumption that $t_\xi \notin F$. Thus $t_\xi \in F$ for all $\xi < \alpha$; show that $p \Vdash (\bigcap_{\xi < \alpha} t_\xi)\hat{} \subseteq \bigcap_{x \in \hat\alpha} f(x)$, and use the κ-completeness of F.)

(e) Show that $V^{(B)} \models \hat\kappa$ *is a measurable cardinal*. (Use (a)-(d).)

Now let ZFM = ZFC + 'there exists a measurable cardinal'.

(ii) Show that, if ZFM is consistent, so is ZFM + $2^{\aleph_0} = \aleph_1$. (Let $P = C_{\omega_1}(\omega_1, P\omega)$ and let $B = B_{\omega_1}(\omega_1, P\omega)$. Notice that P is $\aleph_1$-closed (2.17), so that $V^{(B)} \models (P\omega)\hat{} = P\hat\omega$. Now use 5.1 to show that $V^{(B)} \models |(P\omega)\hat{}| \leq \aleph_1$. Conclude that $V^{(B)} \models 2^{\aleph_0} = \aleph_1$, and use (i) (e).)

(iii) Show that, if ZFM is consistent, so is ZFM $+ 2^{\aleph_0} \geq \aleph_2$. (Use (i) (e) and 2.12.)

Applications to the Theory of Boolean Algebras

It is well-known that any countably generated Boolean algebra is a homomorphic image of the free Boolean algebra on countably many generators. (This algebra may be explicitly described as the algebra of clopen subsets of the Cantor ternary set with its usual topology.) In 1964 Gaifman and Hales (independently) showed that the situation in respect of *complete* Boolean algebras is strikingly different.

Let us say that a complete Boolean algebra is *countably completely generated* (ccg) if there is a countable subset X of B such that the least complete subalgebra of B which includes X is B itself. (Under these conditions X is said to *completely generate B*.) Now Gaifman and Hales showed that there are ccg complete Boolean algebras of *arbitrarily high cardinality*: this implies at once that there is no Boolean algebra B such that each ccg complete Boolean algebra is a homomorphic image of B.

In 1965 Solovay used the properties of collapsing algebras to provide a remarkably simple proof of Gaifman and Hales' theorem. Essentially, Solovay observed that if $B = \mathrm{RO}(\lambda^\omega)$, then $V^{(B)}$ can be obtained by adjoining a B-valued set E of natural numbers to V, and that the Boolean values $[\![\hat{\eta} \in E]\!]$ completely generate B: cf. 4.35 and 4.36.

5.6. Theorem. *Let λ be an infinite cardinal and put $B = \mathrm{RO}(\lambda^\omega)$. Then B is ccg and $|B| \geq \lambda$. Hence there are ccg complete Boolean algebras of arbitrarily high cardinality.*

Proof. Define $f \in V^{(B)}$ by

$$\mathrm{dom}(f) = \{\langle \hat{m}, \hat{\eta} \rangle^{(B)} : \langle m, \eta \rangle \in \omega \times \lambda\}$$

and

$$f(\langle \hat{m}, \hat{\eta} \rangle^{(B)}) = \{g \in \lambda^\omega : g(m) = \eta\}.$$

One can then show, as in the proof of 5.1, that f is a collapsing function from $\hat{\omega}$ to $\hat{\lambda}$ in $V^{(B)}$, i.e.

$$V^{(B)} \models f \text{ is a map of } \hat{\omega} \text{ onto } \hat{\lambda}. \tag{1}$$

Let $b_{m\eta} = f((\hat{m}, \hat{\eta})^{(B)})$; it is then not hard to verify that

$$b_{m\eta} = [\![f(\hat{m}) = \hat{\eta}]\!].$$

The $b_{m\eta}$ form a subbase for the product topology on λ^ω, and using this fact it is straightforward to show that the $b_{m\eta}$ completely generate B. Since there are λ different $b_{m\eta}$, it follows that $|B| \geq \lambda$.

Put $a_{mn} = [\![f(\hat{m}) < f(\hat{n})]\!]$, for $m, n \in \omega$. We shall show that the a_{mn} completely generate B, thereby proving the theorem. Since the $b_{m\eta}$ completely generate B, it suffices to show that each $b_{m\eta}$ is in the complete subalgebra B' of B completely generated by the a_{mn}.

We prove this last assertion by induction on η. Suppose that, for all $\xi < \eta$ and all $m \in \omega$ we have $b_{m\xi} \in B'$. We show that both $[\![f(\hat{m}) < \hat{\eta}]\!]$ and $[\![f(\hat{m}) \leq \hat{\eta}]\!]$ are in B', for all $m \in \omega$. We shall then have

$$b_{m\eta} = [\![f(\hat{m}) = \hat{\eta}]\!] = [\![f(\hat{m}) \leq \hat{\eta}]\!] \wedge [\![f(\hat{m}) < \hat{\eta}]\!]^* \in B',$$

completing the induction step.

First we have

$$[\![f(\hat{m}) < \hat{\eta}]\!] = \bigvee_{\xi < \eta} [\![f(\hat{m}) = \hat{\xi}]\!] = \bigvee_{\xi < \eta} b_{m\xi} \in B', \tag{2}$$

since, by inductive hypothesis, $b_{m\xi} \in B'$ for all $\xi < \eta$. And finally,

$$\begin{aligned}
[\![f(\hat{m}) \leq \hat{\eta}]\!] &= [\![\forall \alpha < f(\hat{m})(\alpha < \hat{\eta})]\!] \\
&= [\![\forall x \in \hat{\omega}[f(x) < f(\hat{m}) \to f(x) < \hat{\eta}]]\!] \quad \text{(using (1))} \\
&= \bigwedge_{n \in \omega} [\![[\![f(\hat{n}) < f(\hat{m})]\!] \Rightarrow [\![f(\hat{n}) < \hat{\eta}]\!]]\!] \\
&= \bigwedge_{n \in \omega} [a_{nm} \Rightarrow [\![f(\hat{n}) < \hat{\eta}]\!]] \in B' \quad \text{(by (2))}. \quad \square
\end{aligned}$$

In 1967 Kripke strengthened Solovay's result by showing that collapsing algebras enjoy a remarkable *embedding* property. If there exists a

5. Cardinal Collapsing

complete monomorphism of a Boolean algebra A into B, let us say that A can be *completely embedded* in B. Then Kripke's result is:

5.7. Theorem. *Let A be a Boolean algebra of infinite cardinality κ. Then A can be completely embedded in the collapsing $(\aleph_0, 2^\kappa)$-algebra.*

Proof. Let $\lambda = 2^\kappa$ and let $B = \mathrm{RO}(\lambda^\omega)$ be the collapsing $(\aleph_0, \lambda)$-algebra. We know from 5.2 that $V^{(B)} \models \hat{\lambda}$ *is countable*, whence

$$V^{(B)} \models (PA)\hat{\,} \text{ is countable}.$$

Let $\{Q_\xi : \xi < \kappa\}$ be a partition of λ and for each $\xi < \kappa$ put

$$b_\xi = \{f \in \lambda^\omega : f(0) \in Q_\xi\}.$$

Then the b_ξ form a partition of unity in B. Let $\{a_\xi : \xi < \kappa\}$ be an enumeration of $A - \{0_A\}$. Then by the Mixing Lemma, there is $b \in V^{(B)}$ such that $b_\xi \leq [\![b = \hat{a}_\xi]\!]$ for all $\xi < \kappa$. It follows at once that

$$[\![b \in \hat{A}]\!] \geq \bigvee_{\xi < \kappa} [\![b = \hat{a}_\xi]\!] \geq \bigvee_{\xi < \kappa} b_\xi = 1.$$

The predicate '*x is a Boolean algebra*' is a restricted formula, so that

$$V^{(B)} \models \hat{A} \text{ is a Boolean algebra}.$$

Moreover we have, for each $\xi < \kappa$,

$$b_\xi \leq [\![b = \hat{a}_\xi]\!] = [\![b = \hat{a}_\xi]\!] \wedge [\![\hat{a}_\xi \neq \hat{0}_A]\!] \leq [\![b \neq \hat{0}_A]\!] = [\![b \neq 0_{\hat{A}}]\!],$$

so that

$$1 = \bigvee_{\xi < \kappa} b_\xi \leq [\![b \neq 0_{\hat{A}}]\!].$$

Let $S = \{X \in PA : \bigvee X \text{ exists in } A\}$. Then $V^{(B)} \models \hat{S}$ *is countable*, and since $V^{(B)} \models$ *Rasiowa-Sikorski Lemma*,

$$V^{(B)} \models \exists U [U \text{ is an } \hat{S}\text{-complete ultrafilter in } \hat{A} \text{ and } b \in U].$$

The Maximum Principle now implies the existence of a $U \in V^{(B)}$ such that

$$V^{(B)} \models U \text{ is an } \hat{S}\text{-complete ultrafilter in } A \text{ and } b \in U. \qquad (1)$$

We define $h: A \to B$ by

$$h(a) = [\![\hat{a} \in U]\!]$$

for $a \in A$. It is easy to verify that h is a homomorphism of A into B. To see that h is complete, observe that, if $X \in S$ and $a = \bigvee X$ in A, then $[\![\hat{a} = \bigvee \hat{X} \text{ in } \hat{A}]\!] = 1$, so that, using (1),

$$\begin{aligned} h(\bigvee X) = h(a) &= [\![\hat{a} \in U]\!] = [\![\bigvee \hat{X} \in U]\!] \\ &= [\![\exists x \in \hat{X}(x \in U)]\!] = \bigvee_{x \in X} [\![\hat{x} \in U]\!] \\ &= \bigvee_{x \in X} h(x) = \bigvee h[X]. \end{aligned}$$

And finally h is one-one, because if $0_A \neq a \in A$, then $a = a_\xi$ for some $\xi < \kappa$, whence $h(a) = [\![\hat{a}_\xi \in U]\!] \geq [\![b \in U]\!] \wedge [\![\hat{a}_\xi = b]\!] = [\![\hat{a}_\xi = b]\!] \neq 0$. □

5.6 and 5.7 immediately give

5.8. Corollary. *Each Boolean algebra can be completely embedded in a ccg complete Boolean algebra.* □

Problems

5.9. (*Universal complete Boolean algebras*). Let κ be an infinite cardinal. A Boolean algebra B is said to be κ-*universal* if for each Boolean algebra A of cardinality $\leq \kappa$ there is a monomorphism of A into B. If B is *complete*, show that the following conditions are equivalent:

(i) B is κ-universal;

(ii) B has an antichain of cardinality κ. (For (ii) $\to$ (i); argue as in the proof of 5.7, ignoring the collapsing property.)

5.10. (*Homogeneous Boolean algebras*). Show that, for each λ, the collapsing $(\aleph_0, \lambda)$-algebra is homogeneous. Deduce that each Boolean algebra can be completely embedded in a homogeneous complete Boolean algebra. (To establish homogeneity, argue along the lines of 3.7).

Chapter 6

Iterated Boolean Extensions, Martin's Axiom and Souslin's Hypothesis

Souslin's Hypothesis

It is well known that the real line can be characterized up to order isomorphism as the unique linearly ordered set which is order dense, complete and unbounded, and *separable, i.e.* has a countable subset which meets each non-empty open interval. In 1920 Souslin raised the question as to whether separability could be replaced by the following—apparently weaker—condition:

$$\text{Every family of disjoint open intervals is countable}. \quad (*)$$

Souslin's problem may be equivalently stated in the form of what is usually called

Souslin's Hypothesis (SH). *Every order dense linearly ordered set satisfying* $(*)$ *above is separable.*

To see that the two formulations are equivalent, let us write SH′ for the assertion that any complete, unbounded, order dense, linearly ordered set satisfying $(*)$ is isomorphic to the real line R. Then clearly SH → SH′. Conversely, if SH′ holds, let P be an (infinite) order dense linearly ordered set satisfying $(*)$. Then P contains a copy of the ordered set Q of rational numbers. The (Dedekind) order completion C of P (with its end-points removed) is then order dense, unbounded, and satisfies $(*)$ since P does.

Thus SH' implies that C is isomorphic to R. So P may be regarded as a dense subset of R which contains Q. Clearly P is then separable, and SH follows.

In this chapter we are going to establish first the independence and then the relative consistency of SH with ZFC. The first step in this process is to introduce the notion of a tree.

A *tree* is a partially ordered set $\langle T, \leq_T \rangle$ with the property that, for each $x \in T$, the set $\{y : y <_T x\}$ of predecessors of x is well ordered by $\leq_T$. For each $x \in T$, we write $o(x)$ for the order type of $\{y : y <_T x\}$; $o(x)$ is then an ordinal. For each ordinal α, the αth *level* of T consists of all $x \in T$ for which $o(x) = \alpha$. The *height* of T is the least α such that the αth level of T is empty. A *branch* in T is a maximal linearly ordered subset of T. A subset X of T is said to be *free* if any two different elements x, y of X are incomparable, i.e., neither $x \leq_T y$ nor $y \leq_T x$. Finally, a tree T is called a *Souslin tree* if

T has height ω_1;
every branch in T is countable;
every free subset of T is countable.

We can now reformulate Souslin's hypothesis in terms of Souslin trees.

6.1. Lemma. SH *holds iff there are no Souslin trees.*

Proof. Sufficiency. Suppose SH fails; then there is a dense linearly ordered set P which is not separable but in which every family of disjoint open intervals is countable. We use P to construct a Souslin tree T as follows. T will consist of closed (nondegenerate) intervals in P, and will be partially ordered by *inverse* inclusion $\supseteq$.

We construct T by recursion on $\alpha < \omega_1$. Let $I_0 = [a_0, b_0]$ be arbitrary (with $a_0 < b_0$). Assuming that we have got all I_β for $\beta < \alpha$, consider the countable set $C = \{a_\beta : \beta < \alpha\} \cup \{b_\beta : \beta < \alpha\}$ of endpoints of the intervals I_β. Since P is not separable, there must be an interval disjoint from C; let $I_\alpha = [a_\alpha, b_\alpha]$ be one. The set $T = \{I_\alpha : \alpha < \omega_1\}$ is

6. Iterated Boolean Extensions

uncountable and partially ordered by $\supseteq$. If $\alpha < \beta$, then either $I_\alpha \supseteq I_\beta$ or $I_\alpha \cap I_\beta = \emptyset$. It follows that, for each α, the set $\{I \in T : I \supseteq I_\alpha\}$ is well-ordered by $\supseteq$ and so T is a tree.

We show that T has no uncountable branches and no uncountable free subsets. Clearly the height of T then cannot exceed ω_1; and since every level of T is evidently free and T is uncountable, it follows that T has height ω_1.

If I, J are incomparable members of T, then they are, by construction, disjoint intervals of P; so any free subset of T is countable. To show that T has no uncountable branches, we observe that if b is a branch of length ω_1, then the left endpoints of the intervals $I \in b$ form an increasing sequence $\{x_\alpha : \alpha < \omega_1\}$ of points of P. But then the intervals $(x_\alpha, x_{\alpha+1})$, $\alpha < \omega_1$ form an uncountable collection of disjoint open intervals in P, contradicting assumption.

Necessity. Let T be a Souslin tree. First let us remove from T all points $x \in T$ such that $\{y \in T : x \leq y\}$ is countable, thus obtaining a new tree T'. It is easy to see that for each $x \in T'$ there exists $y \in T'$, $y > x$, at each greater level $< \omega_1$. Next, we discard from T' all points $x \in T'$ for which there is only *one* point $y > x$ at the next level. This gives us a new tree T''. Finally, from T'' we expunge all points except those at *limit* levels. This yields a Souslin tree T''' in which each level has cardinality $\aleph_0$. Thus, without loss of generality we may assume that the same holds for T.

Now let P be the set of all branches in T; we order P as follows. First, we order each level of T as in the rational numbers. Then, given $b_1, b_2 \in P$, we put $b_1 < b_2$ if the αth element of b_1 precedes the αth element of b_2 in the ordering of U_α, where U_α is the least level at which b_1 and b_2 differ. Clearly this prescription makes P linearly ordered and dense.

If (a, b) is an interval in P, it is not hard to see that there is $x \in T$ such that $I_x \subseteq (a, b)$ where I_x is the interval $I_x = \{c \in P : x \in c\}$. Moreover, if $I_x \cap I_y = \emptyset$, then x and y are incomparable points of T. It follows that every collection of disjoint open intervals in P is countable.

On the other hand, P is not separable. For if C is a countable set of branches in T, let α be a countable ordinal exceeding the length of any branch in C. Then if x is any point at level α, the interval I_x is disjoint from C. □

The Independence of SH

Before we begin the proof of independence of SH from ZFC, we require a combinatorial lemma.

6.2. Lemma. *Let S be an uncountable collection of finite sets. Then there is an uncountable $Z \subseteq S$ and a finite set A such that $X \cap Y = A$ for any distinct elements $X, Y \in Z$.*

Proof. Since S is uncountable, it is clear that uncountably many $X \in S$ have the same cardinality. Thus we may assume that, for some n, $|X| = n$ for all $X \in S$. The lemma is now proved by induction on n. If $n = 1$, the lemma is trivial. So assume its truth for n, and let S be such that $|X| = n + 1$ for all $X \in S$.

Case 1. Some element a belongs to uncountably many $X \in S$. In this case we obtain the required $Z \subseteq S$ by applying the inductive hypothesis to the family $\{X - \{a\} : X \in S \wedge a \in X\}$.

Case 2. Not case 1. Here we can easily construct a disjoint family $Z = \{X_\alpha : \alpha < \omega_1\} \subseteq S$ by choosing inductively X_α to be a member of S disjoint from all X_ξ for $\xi < \alpha$. □

Now let M be any countable transitive $\in$-model of ZFC. We shall show that M has a generic extension $M[G]$ which contains a Souslin tree.

In M, let P be the set of all finite trees $\langle T, \leq_T \rangle$ such that $T \subseteq \omega_1$ and $\alpha \leq_T \beta \to \alpha \leq \beta$ (as ordinals). We partially order P by stipulating that:

$$\langle T_1, \leq_{T_1} \rangle \preceq \langle T_2, \leq_{T_2} \rangle \leftrightarrow T_1 \supseteq T_2 \wedge \leq_{T_2} = \leq_{T_1} | T_2.$$

In the sequel we shall usually write "$\leq_1$" for "$\leq_{T_1}$", etc.

A partially ordered set is said to satisfy the *countable chain condition* (ccc) if every subset consisting of incompatible elements is countable. It is

6. Iterated Boolean Extensions

clear that a partially ordered set satisfies ccc in this sense iff its Boolean completion (cf. 2.4) satisfies the ccc in the sense of Chapter 1.

6.3. Lemma. $\langle P, \preceq \rangle$ *satisfies* ccc.

Proof. Let S be an uncountable subset of P. Using 6.2, we obtain an uncountable subset $Z_1 \subseteq S$ and a finite set $A \subseteq \omega_1$ such that, for any distinct $T_1, T_2 \in Z_1$ we have $T_1 \cap T_2 = A$ and $\leq_1 |A = \leq_2|A$. Now discard from Z_1 all trees T for which there exist $\alpha \in A$ and $\beta < \alpha$ such that $\beta \in T - A$. Only countably many trees are lost from Z_1 in this way. If we call what is left Z_2, then Z_2 is uncountable. But now any $T_1, T_2 \in Z$ are compatible: for if we define $\leq_3$ on $T_3 = T_1 \cup T_2$ by

$$\alpha \leq_3 \beta \leftrightarrow \alpha \leq_1 \beta \vee \alpha \leq_2 \beta,$$

then $\langle T_3, \leq_3 \rangle \preceq \langle T_1, \leq_1 \rangle$ and $\langle T_3, \leq_3 \rangle \preceq \langle T_2, \leq_2 \rangle$. The lemma follows. □

Now, using 4.34, we choose an M-generic subset G of P and put $H = \bigcup G$. We let $\leq_H$ be the partial ordering on H induced in the obvious way by the partial orderings of the members of G. It is easy to see that H is a tree in $M[G]$.

6.4. Theorem. $M[G] \models H$ *is a Souslin tree.*

Proof. Let us call a point x of a tree a *branch point* if there are at least two points $y > x$ at the next level above x.

We claim first that:

(1) every point of H is a branch point;

(2) for each $\alpha \in H$, the set $\{\beta \in H : \alpha <_H \beta\}$ is uncountable in $M[G]$.

To prove (1), we take any $\alpha \in H$, choose $T_0 \in G$ such that $\alpha \in T_0$ and observe that the set of $T \in P$ in which α is a branch point is dense below T_0 in P. Consequently there must be $T \in G$ in which α is a branch point; it follows that α is a branch point in H.

To prove (2), we again take $\alpha \in H$ and $T_0 \in G$ such that $\alpha \in T_0$.

Next we observe that, for each $\xi < \omega_1^{(M)}$, the set

$$Q = \{T \in P : \alpha \in T \wedge \exists \beta \in T[\xi < \beta \wedge \alpha <_T \beta]\}$$

is dense below T_0. Since G is generic, $G \cap Q \neq \emptyset$. Hence for each $\xi < \omega_1^{(M)}$ there is $\beta \in H$ such that $\xi < \beta$ and $\alpha <_H \beta$. But 6.3 implies that $\omega_1^{(M)} = \omega_1^{(M[G])}$, and (2) follows.

Now we can prove

(3) $M[G] \models$ *every free subset of H is countable*.

For suppose (3) is false; then, for some $E \in M[G]$ we have

E is an uncountable free subset of H.

Let B be the Boolean completion of P, let i be the canonical map of $M^{(B)}$ onto $M[G]$, and let $\tilde{E}, \tilde{H} \in M^{(B)}$ be such that $i(\tilde{E}) = E$, $i(\tilde{H}) = H$. Then, for some $T_0 \in G$ we have

$$T_0 \Vdash \tilde{E} \text{ is an uncountable free subset of } \tilde{H}. \qquad (*)$$

It follows that for each $\xi < \omega_1^{(M)}$ we can find $T \in P$ and $\alpha_T \in T$ such that $T \preceq T_0$, $\xi < \alpha_T$ and $T \Vdash \hat{\alpha}_T \in \tilde{E}$. In this way we obtain an uncountable (in M) set $S \subseteq P$ and, for each $T \in S$ an ordinal $\alpha_T \in T$ such that $T \Vdash \hat{\alpha}_T \in \tilde{E}$. Lemma 6.2 now yields an uncountable subset Z_1 of S and a finite set A such that, for any $T_1 \neq T_2$ in Z_1 we have $T_1 \cap T_2 = A$ and $\leq_{T_1} |A = <_{T_2}|A$. Without loss of generality we may assume that $\alpha_T \notin A$ for any $T \in Z_1$. Since A is finite, there must be an uncountable subset Z_2 of Z such that, for any $T_1, T_2 \in Z_2$,

$$A \cap \{\beta : \beta <_{T_1} \alpha_{T_1}\} = A \cap \{\beta : \beta <_{T_2} \alpha_{T_2}\}.$$

From Z_2 we discard all trees T for which there is $\alpha \in A$ and $\beta < \alpha$ such that $\beta \in T - A$. Only countably many trees are lost in this way, so if we call what is left Z_3, then Z_3 is uncountable.

If $T_1, T_2 \in Z_3$ then we have $\langle T_3, \leq_3 \rangle \preceq \langle T_1, \leq_1 \rangle$ and $\langle T_3, \leq_3 \rangle \preceq \langle T_2, \leq_2 \rangle$ where $T_3 = T_1 \cup T_2$ and $\leq_3$ is defined as follows: if $\alpha_{T_1} < \alpha_{T_2}$, then

$$\alpha \leq_3 \beta \leftrightarrow \alpha \leq_1 \beta \vee \alpha \leq_2 \beta$$
$$\vee \, [\alpha \leq_1 \alpha_{T_1} \wedge \exists \gamma \leq_2 \alpha_{T_2}(\alpha \leq \gamma \wedge \gamma \leq_2 \beta)]$$
$$\vee \, [\alpha \leq_2 \alpha_{T_2} \wedge \exists \gamma \leq_1 \alpha_{T_1}(\alpha \leq \gamma \wedge \gamma \leq_1 \beta)]$$
$$\vee \, [\alpha \leq_1 \alpha_{T_1} \wedge \alpha_{T_2} \leq_2 \beta],$$

and similarly when $\alpha_{T_2} < \alpha_{T_1}$.

It is clear that α_{T_1} and α_{T_2} are comparable with respect to $\leq_{T_3}$. So

$$T_3 \Vdash \hat{\alpha}_{T_1} \in \tilde{E} \wedge \hat{\alpha}_{T_2} \in \tilde{E} \wedge \hat{\alpha}_{T_1} \text{ is comparable in } \tilde{H} \text{ with } \hat{\alpha}_{T_2}.$$

But this contradicts (∗). This proves (3).

It therefore remains to prove

(4) $M[C] \models H$ *has height* ω_1;

(5) $M[G] \models$ *every branch in H is countable.*

To prove (4), we suppose that (in $M[G]$) H has height $< \omega_1$. Then for any $\alpha \in H$ the set $\{o(\beta) : \alpha <_H \beta\}$ is countable and it follows from (2) that for some $\gamma < \omega_1$ the set $\{\beta \in H : o(\beta) = \alpha\}$ is uncountable. But this latter set is free, contradicting (3).

Finally, suppose (5) is false; let b be a branch in H of length ω_1 (in $M[G]$). Using (1) we choose for each $x \in b$ an element $f(x) \geq_H x$ not in b. Then $\{f(x) : x \in b\}$ is free and uncountable in $M[G]$, contradicting (3). □

6.5. Corollary. *If ZF is consistent, so are* ZFC + GCH + ¬ SH *and* ZFC + ¬ CH + ¬ SH.

Proof. Let B be the Boolean completion of P, and assume that $M \models$ GCH. Since P satisfies ccc, so does B and it is easily shown, using GCH in M, that $M \models |B| = 2^{\aleph_0}$. It follows from 2.8 and 6.4 that $M[G] \models$ GCH ∧ ¬ SH. Since this holds for an arbitrary generic set G in P, we infer that $M[U] \models$ GCH ∧ ¬ SH for an arbitrary generic ultrafilter U in B and hence, using 4.26, that $M^{(B)} \models$ GCH ∧ ¬ SH. By the remarks following 4.25, we can transfer this to $V^{(B)}$, obtaining (now under the assumption that GCH holds in V) that $V^{(B)} \models$ GCH ∧ ¬ SH. The first assertion now follows from 1.19.

If on the other hand $M \models 2^{\aleph_0} \geq \aleph_2$, then $M[G] \models \aleph_2^{(M)} \leq 2^{\aleph_0}$. But B satisfies ccc, and therefore, by 1.51, we have $M[G] \models \aleph_2^{(M)} = \aleph_2$. Hence $M[G] \models \aleph_2 \leq 2^{\aleph_0}$, yielding the second assertion as above. □

Martin's Axiom

Having established the independence of SH from ZFC, we want now to demonstrate its relative consistency. Rather than attempting to go about doing this directly, however, we instead formulate a principle that easily implies the non-existence of Souslin trees, and give a relative consistency proof for this principle. The principle, known as *Martin's axiom*, provides an interesting alternative to the continuum hypothesis, and has become an important tool in general topology and infinite combinatorics.

Let κ be an infinite cardinal. *Martin's axiom at level κ* is the assertion

MA_κ : *If B is a Boolean algebra satisfying ccc and S is any family of subsets of B (each member of which has a join) such that $|S| < \kappa$, then there is an S-complete ultrafilter in B.*

Notice that $\mathrm{MA}_{\aleph_1}$ is just a special case of the Rasiowa-Sikorski lemma, and hence provable in ZFC. So MA_κ is only interesting when $\kappa > \aleph_1$. We observe, however that

6.6. Lemma.

$$\mathrm{ZFC} \vdash \mathrm{MA}_\kappa \to \kappa \leq 2^{\aleph_0}.$$

Proof. Let $B = \mathrm{RO}(2^\omega)$; then B satisfies ccc. For each $s \subseteq \omega$ let χ_s be the characteristic function of s and let $X_s = \{N(p) : p \not\subseteq \chi_s\}$. It is easily verified that $\bigvee X_s = 1$ in B; so if MA_κ with $\kappa > 2^{\aleph_0}$ there would be a $\{X_s : s \subseteq \omega\}$-complete ultrafilter U in B. Now define $t \subseteq \omega$ by

$$n \in t \leftrightarrow N(\{\langle n, 1\rangle\}) \in U.$$

6. Iterated Boolean Extensions

It is now easy to show that for any subset $s \subseteq \omega$ we have $t \neq s$, which is a contradiction. □

6.7. Problem. (*A stronger form of Martin's axiom?*). Let $\mathrm{MA}_\kappa!$ be obtained from MA_κ by dropping the restriction to algebras satisfying ccc. Show that $\mathrm{MA}_\kappa! \to \kappa \leq \aleph_1$. (Consider the collapsing $(\aleph_0, \aleph_1)$-algebra.)

We next give several alternative formulations of MA_κ. If P is a partially ordered set, and S a family of subsets of P, a subset G of P is said to be *S-generic* (cf. 4.34) if

(a) $x \in G, x \leq y \to y \in G$;

(b) $x, y \in G \to \exists z \in G[z \leq x \wedge z \leq y]$;

(c) $X \in S \wedge X$ dense in $P \to X \cap G \neq \emptyset$.

6.8. Theorem. *The following are equivalent for any infinite cardinal κ:*

(i) MA_κ.

(ii) *If B is a Boolean algebra satisfying ccc with $|B| < \kappa$ and S is a family of subsets with $|S| < \kappa$, each member of which has a join, then there is an S-complete ultrafilter in B.*

(iii) *If P is a partially ordered set satisfying ccc such that $|P| < \kappa$ and S is a family of subsets of P with $|S| < \kappa$, then there is an S-generic subset of P.*

(iv) *Same as (iii) but with "$|S| < \kappa$" omitted.*

Proof. (i) $\to$ (ii) is trivial.

(ii) $\to$ (iii). Assume (ii) and let P be a partially ordered set satisfying ccc with $|P| < \kappa$. Let C be the Boolean completion of P and let $j: P \to C$ be the canonical map. Also let B be the subalgebra of C generated by $j[P]$. Then $|B| < \kappa$. Now observe that if E is any family of $< \kappa$ dense subsets of B, then $\bigvee X = 1$ in B for each $X \in E$, and therefore, by (ii), there is an E-complete ultrafilter in B. If S is any family of dense sets in P with $|S| < \kappa$, consider the following dense subsets of B: (a) all $j[X]$

with $X \in S$; (b) all $j[Z_{xy}]$ for $x,y \in P$ where

$$Z_{xy} = \{z \in P : [z \leq x \wedge z \leq y] \vee \neg \operatorname{Comp}(z,x) \vee \neg \operatorname{Comp}(z,y)\}.$$

Since $|P| < \kappa$, there are $< \kappa$ sets of the form (a) or (b); so (ii) yields an ultrafilter U in B which meets all of them. It is now readily verified that $j^{-1}[U]$ is an S-generic subset of P.

(iii) $\to$ (iv). Assume (iii) and let P be a partially ordered set satisfying ccc. Let S be a family of $< \kappa$ dense subsets of P. For each $X \in S$, let Q_X be a maximal subset of X consisting of mutually incompatible elements. Since P satisfies ccc, each Q_X is countable. Hence there is a subset Q of P of cardinality $< \kappa$ such that $Q_X \subseteq Q$ for all $X \in S$ and

$$x,y \in Q \wedge \operatorname{Comp}(x,y) \to \exists z \in Q[z \leq x \wedge z \leq y].$$

Each Q_X is a maximal incompatible subset of Q and, for each $X \in S$, $K_X = \{x \in Q : x \leq y \text{ for some } y \in Q_X\}$ is dense in Q.

The partially ordered set Q is of cardinality $< \kappa$ and satisfies ccc. So by (iii) there is a $\{K_X : X \in S\}$-generic subset G of Q. It is now easily verified that the set $\{x \in P : \exists y \in G[y \leq x]\}$ is S-generic in P.

(iv) $\to$ (i). Assume (iv), let B be a Boolean algebra with ccc, let S be a family of subsets of B, each member of which has a join, let $P = B - \{0\}$ and for each $X \in S$ let

$$D_X = \{y \in P : \exists x \in X[y \leq x] \vee \forall x \in X[y \wedge x = 0]\}.$$

Each D_X is dense in P and so (iv) implies that there is a $\{D_X : X \in S\}$-generic subset G of P. Clearly any ultrafilter in B extending G is S-complete. □

The principle $\operatorname{MA}_{2^{\aleph_0}}$ is called *Martin's axiom* and is written simply MA. Clearly MA is a consequence of CH; but we shall see that MA can hold even when CH fails. Moreover, we can use 6.8 to show that, in this eventuality, Souslin's hypothesis holds:

6.9. Theorem.

$$\operatorname{MA} + 2^{\aleph_0} > \aleph_1 \to \operatorname{SH}.$$

6. Iterated Boolean Extensions

Proof. Suppose SH is false; then there is a Souslin tree T. Let T' be the set of $x \in T$ for which $\{y \in T : x \leq_T y\}$ is uncountable. It is then easy to see that T' is a Souslin tree with the property:

for each $x \in T'$ there is some $y > x$ at each greater level $(< \omega_1)$. $\quad(*)$

Now let P be the partially ordered set obtained from T' by reversing the order. It is easy to see that P satisfies ccc. For each $\alpha < \omega_1$ let

$$X_\alpha = \{x \in T' : o(x) > \alpha\}.$$

Using $(*)$, one verifies that X_α is dense in P.

Thus, assuming $MA + 2^{\aleph_0} > \aleph_1$, there is an $\{X_\alpha : \alpha < \omega_1\}$-generic subset G of P. It is now a routine matter to verify that G is a branch in T' of cardinality ω_1, a contradiction. $\square$

It follows from 6.9 that, in order to establish the relative consistency of SH, it suffices to establish that of $MA + 2^{\aleph_0} > \aleph_1$. We shall achieve this by constructing an increasing sequence of Boolean extensions of the universe in such a way that at each succesive stage a potential counterexample to Martin's axiom is "liquidated" by adjoining an ultrafilter of the appropriate sort. Then, with some finesse, we can show that the "limit" of this sequence of Boolean extensions (in a sense to be made precise later on) is a Boolean-valued model of $MA + 2^{\aleph_0} > \aleph_1$. We turn now to elaborating this procedure, which is called the method of *iterated Boolean extensions*.

Iterated Boolean Extensions

Let B be a complete Boolean algebra, and suppose we are given elements $C, \leq_C$ of $V^{(B)}$ such that

$$V^{(B)} \models \langle C, \leq_C \rangle \text{ is a Boolean algebra }.$$

Let A be a *core* for C (see Chapter 1). We shall see that A carries the structure of a Boolean algebra in a natural way.

First we define a relation $\leq_A$ on A by putting, for each $a, a' \in A$,

$$a \leq_A a' \leftrightarrow [\![a \leq_C a']\!] = 1$$

(cf. proof of 1.43). It is easily verified that $\leq_A$ is a partial ordering on A and that with this partial ordering A is a Boolean algebra in which the Boolean operations $\wedge_A$, $\vee_A$, $*_A$ are given by

$$a \wedge_A a' = \text{ unique } x \in A \text{ for which } [\![x = a \wedge_C a']\!] = 1;$$
$$a \vee_A a' = \text{ unique } x \in A \text{ for which } [\![x = a \vee_C a']\!] = 1;$$
$$a^{*_A} = \text{ unique } x \in A \text{ for which } [\![x = a^{*_C}]\!] = 1,$$

where $\wedge_C$, $\vee_C$, $*_C$ are the Boolean operations in C (in $V^{(B)}$). It is also not hard to see that, despite the apparent freedom in the choice of the core A, the Boolean algebra $\langle A, \leq_A \rangle$ is determined uniquely up to isomorphism. *We shall write $B \otimes C$ for A and $\leq_{B \otimes C}$ or $\leq$ for $\leq_A$.*

If $x, y \in B \otimes C$ and $b \in B$, then the 2-term mixture $b.x + b^*.y$ is, with probability 1, an element of C, and hence there is a unique element of $B \otimes C$ which is equal to it with probability 1. Without loss of generality we may and shall assume in the sequel that this element of $B \otimes C$ is $b.x + b^*.y$. That is, *we shall assume that $B \otimes C$ is closed under two-term mixtures of the form $b.x + b^*.y$.* This fact should be borne in mind when considering the next problem.

6.10. Problem. (*An isomorphism of Boolean algebras*). Show that the map $p: B \to B \otimes \hat{2}$ defined by $p(b) = b.\hat{1} + b^*.\hat{0}$ is an isomorphism of Boolean algebras.

Next, we treat the computation of arbitrary joins in $B \otimes C$.

6.11. Lemma. *Let $X \subseteq B \otimes C$ and put $X' = X \times \{1\}$. Then $X' \in V^{(B)}$ and $[\![X' \subseteq C]\!] = 1$. If $a \in B \otimes C$ satisfies $[\![a = \bigvee_C X']\!] = 1$, then $a = \bigvee X$ in $B \otimes C$.*

Proof. If $x \in X$, then clearly $[\![x \in X']\!] = 1$, so that $[\![x \leq_C a]\!] = 1$, whence $x \leq_{B \otimes C} a$. Therefore a is an upper bound for X in $B \otimes C$. Also, if $y \in B \otimes C$ is any upper bound for X, then $[\![x \leq_C y]\!] = 1$ for all $x \in X$, whence $[\![y \text{ is an upper bound for } X']\!] = 1$, so that $[\![a \leq_C y]\!] = 1$, and therefore $a \leq_{B \otimes C} y$. So $a = \bigvee X$ in $B \otimes C$ as claimed. □

6. Iterated Boolean Extensions

As an immediate consequence we have the

6.12. Corollary. *If*

$$V^{(B)} \models \langle C, \leq_C \rangle \text{ is a complete Boolean algebra},$$

then $B \otimes C$ is complete. □

Next, we show that B is completely embeddable in $B \otimes C$. In $V^{(B)}$ we have the natural monomorphism i of the two element Boolean algebra $\hat{2}$ into C which sends $\hat{0}$ to 0_C and $\hat{1}$ to 1_C, where 0_C, 1_C are the unique elements of $B \otimes C$ which with probability 1 are the least and largest elements of C respectively. This in turn induces the natural map $j \colon B \otimes \hat{2} \to B \otimes C$ defined by setting

$$j(x) = \text{ unique } y \in B \otimes C \text{ for which } [\![y = i(x)]\!] = 1.$$

Clearly, for $x \in B \otimes \hat{2}$,

$$[\![x = \hat{1}]\!] = [\![j(x) = 1_C]\!];$$
$$[\![x = \hat{0}]\!] = [\![j(x) = 0_C]\!],$$

and so $j(x)$ can be described as the two-term mixture

$$j(x) = [\![x = \hat{1}]\!].1_C + [\![x = \hat{0}]\!].0_C$$

for $x \in B \otimes \hat{2}$. By 6.10, we have a natural isomorphism $p \colon B \cong B \otimes \hat{2}$ given by $p(b) = b.\hat{1} + b^*.\hat{0}$ for $b \in B$. Thus the composition $e = j \circ p$ is given by

$$e(b) = j(p(b)) = [\![p(b) = \hat{1}]\!].1_C + [\![p(b) = \hat{0}]\!].0_C = b.1_C + b^*.0_C.$$

The map $e \colon B \to B \otimes C$ is then given by

$$\text{for } b \in B, e(b) = \text{ the unique } x \in B \otimes C \text{ for which} \quad (6.13)$$
$$[\![x = 1_C]\!] = b, [\![x = 0_C]\!] = b^*.$$

Observe that by the very definition of e we have for $b \in B$

$$V^{(B)} \models e(b) \in \{0_C, 1_C\}. \quad (6.14)$$

Recall that a homomorphism h of Boolean algebras is said to be *complete* if it preserves arbitrary joins, that is, if whenever X is a subset of the domain algebra of h such that $\bigvee X$ exists, then $\bigvee h[X]$ exists in the codomain algebra of h and is equal to $h(\bigvee X)$.

6.15. Lemma. *The map e is a complete monomorphism of B into $B \otimes C$.*

Proof. The fact that e is injective and preserves complements is easily established and is left to the reader. We show that e preserves arbitrary joins in B. Suppose then that $X \subseteq B$. Let $Y = e[X]$ and $Y' = Y \times \{1\}$. Then $[\![Y' \subseteq C]\!] = 1$ and, if we choose $a \in B \otimes C$ to satisfy $[\![a = \bigvee_C Y']\!] = 1$, then by 6.11 we have $a = \bigvee Y$ in $B \otimes C$. Now (6.14) gives $[\![Y' \subseteq \{0_C, 1_C\}]\!] = 1$, so that $[\![a \in \{0_C, 1_C\}]\!] = 1$, whence

$$[\![a = 1_C]\!] = [\![\bigvee_C Y' = 1_C \wedge Y' \subseteq \{0_C, 1_C\}]\!]$$
$$= [\![1_C \in Y']\!]$$
$$= \bigvee_{x \in X} [\![e(x) = 1_C]\!]$$
$$= \bigvee X.$$

Therefore a satisfies the defining equations (6.13) for $e(\bigvee X)$ in $B \otimes C$; in other words $e(\bigvee X) = \bigvee e[X]$. Thus e is complete as claimed. □

In view of 6.15, e is called the *canonical embedding* of B in $B \otimes C$.

We continue with some technical lemmas which we shall require later on.

6.16. Lemma. *For $x, y \in B \otimes C$, $b \in B$, we have*

$$b \leq [\![x \leq_C y]\!] \leftrightarrow e(b) \wedge x \leq_{B \otimes C} y.$$

Proof. We have

$$b = [\![e(b) = 1_C]\!].$$

So

$$b \leq [\![x \leq_C y]\!] \leftrightarrow V^{(B)} \models [e(b) = 1_C \to x \leq_C y]. \qquad (*)$$

6. Iterated Boolean Extensions

But by (6.14)

$$V^{(B)} \models e(b) = 1_C \vee e(b) = 0_C.$$

So

$$V^{(B)} \models e(b) = 1_C \rightarrow x \leq_C y \leftrightarrow V^{(B)} \models e(b) \wedge x \leq_C y$$
$$\leftrightarrow e(b) \wedge x \leq_{B \otimes C} y.$$

The result now follows from (∗). □

6.17. Lemma. *Let* $X \in V^{(B)}$. *Then there is* $Y \in V^{(B)}$ *such that* $\mathrm{dom}(Y) \subseteq B \otimes C, Y$ *is definite (1.40) and*

$$[\![X \subseteq C]\!] \leq [\![Y = X \cup \{0_C\}]\!].$$

Proof. Put $b = [\![X \subseteq C]\!]$. Using the Mixing Lemma, choose $X' \in V^{(B)}$ to satisfy

$$[\![X' = X \cup \{0_C\}]\!] \geq b,$$
$$[\![X' = \{0_C\}]\!] \geq b^*.$$

Then $[\![\emptyset \neq X' \subseteq C]\!] = 1$. Now put

$$Y' = \{y \in B \otimes C : [\![y \in X']\!] = 1\}$$

and $Y = Y' \times \{1\}$. Notice that Y' is a core for X'.

We claim that Y meets the required conditions; to establish this it clearly suffices to show that $[\![Y = X']\!] = 1$.

Firstly, we have, for any $x \in V^{(B)}$,

$$[\![x \in Y]\!] = \bigvee_{y \in Y'} [\![x = y]\!]$$
$$= \bigvee_{y \in Y'} [\![x = y]\!] \wedge [\![y \in X']\!]$$
$$\leq [\![x \in X']\!],$$

so $[\![Y \subseteq X']\!] = 1$. Moreover, since $[\![X' \neq \emptyset]\!] = 1$, by 1.32 there is, for each $x \in V^{(B)}$, an $x' \in Y'$ such that $[\![x = x']\!] = [\![x \in X']\!]$. Hence

$$\begin{aligned} [\![x \in X']\!] &= [\![x = x']\!] \\ &\leq \bigvee_{y \in Y'} [\![x = y]\!] \\ &= [\![x \in Y]\!], \end{aligned}$$

so $[\![X' \subseteq Y]\!] = 1$. Thus the claim, and the lemma, is proved. $\square$

We recall from 3.12(i) that the complete monomorphism $e \colon B \to B \otimes C$ induces a natural map $\bar{e} \colon V^{(B)} \to V^{(B \otimes C)}$. We shall *identify B with its image in $B \otimes C$, so that B becomes identified as a complete subalgebra of $B \otimes C$ and $V^{(B)}$ as a subclass of $V^{(B \otimes C)}$*. (Notice that this amounts to taking e, and hence $\bar{e}$, as the identity map.)

Since

$$V^{(B)} \models C \text{ is a } P^{(B)}(C)\text{-complete Boolean algebra,}$$

it follows from 1.21 that

$$V^{(B \otimes C)} \models C \text{ is a } P^{(B)}(C)\text{-complete Boolean algebra.}$$

We are going to show that, in $V^{(B \otimes C)}$, C contains a $P^{(B)}(C)$-complete ultrafilter.

Define the object $U^+ \in V^{(B \otimes C)}$ by $\mathrm{dom}(U^+) = B \otimes C$ and $U^+(a) = a$ for all $a \in B \otimes C$. That is, U^+ is the identity map on $B \otimes C$. Then, for $a \in B \otimes C$, we have

$$[\![a \in U^+]\!] = a. \tag{6.18}$$

To prove this, we note first that since the map e is now the identity, Lemma 6.16 becomes, for $x, y \in B \otimes C$, $b \in B$,

$$b \leq [\![x \leq_C y]\!] \leftrightarrow b \wedge x \leq_{B \otimes C} y. \tag{6.19}$$

Now $[\![x = a]\!] \leq [\![x \leq_C a]\!]$, so taking $b = [\![x = a]\!]$ we get from 6.16,

$$[\![x = a]\!] \wedge x \leq_{B \otimes C} a.$$

6. Iterated Boolean Extensions

Thus

$$[\![a \in U^+]\!] = \bigvee_{x \in B \otimes C} [\![x = a]\!] \wedge x \leq a$$

and the reverse inequality follows from 1.17(ii).

Our next result is crucial.

6.20. Theorem.

$V^{(B \otimes C)} \models U^+$ *is a* $P^{(B)}(C)$-*complete ultrafilter in* C.

Proof. The proof proceeds somewhat along the lines of 4.21 (of which the present theorem is actually a generalization), only it is a little more troublesome. We shall only verify two of the properties that U^+ must have, leaving the verification of the others to the reader.

(a) $[\![\forall xy \in U^+[x \wedge_C y \in U^+]]\!] = 1$. To verify this, observe that the l.h.s. is:

$$\bigwedge_{a,b \in B \otimes C} [U^+(a) \wedge U^+(b)] \Rightarrow [\![a \wedge_C b \in U^+]\!] = \bigwedge_{a,b \in B \otimes C} [a \wedge b \Rightarrow a \wedge b] = 1.$$

(b) $[\![\forall X \in P^{(B)}(C)[\bigvee_C X \in U^+ \to U^+ \cap X \neq \emptyset]]\!] = 1$. To verify this, notice that the l.h.s. is

$$\bigwedge_{X \in \mathrm{dom}(P^{(B)}(C))} [\![X \subseteq C \wedge \bigvee_C X \in U^+]\!] \Rightarrow [\![U^+ \cap X \neq \emptyset]\!].$$

Given $X \in \mathrm{dom}(P^{(B)}(C))$, take $Y \in V^{(B)}$ to satisfy the conditions of 6.17. Then

$$[\![X \subseteq C \wedge \bigvee_C X \in U^+]\!] \leq [\![\bigvee_C Y \in U^+]\!].$$

Now, by 6.11 there is an $a \in B \otimes C$ such that $[\![a = \bigvee_C Y]\!] = 1$ and also such that $a = \bigvee_{B \otimes C} \mathrm{dom}(Y)$. Therefore

$$[\![\bigvee_C Y \in U^+]\!] = [\![a \in U^+]\!] = a = \bigvee_{B \otimes C} \mathrm{dom}(Y).$$

Hence

$$[\![X \subseteq C \land \bigvee_C X \in U^+]\!] \le \bigvee_{B \otimes C} \mathrm{dom}(Y)$$
$$= \bigvee_{x \in \mathrm{dom}(Y)} [\![x \in U^+]\!]$$
$$= [\![\exists x \in Y[x \in U^+]]\!]$$
$$= [\![U^+ \cap Y \ne \emptyset]\!].$$

So

$$[\![X \subseteq C \land \bigvee_C X \in U^+]\!] \le [\![U^+ \cap Y \ne \emptyset]\!] \land [\![X \subseteq C]\!]$$
$$\le [\![U^+ \cap Y \ne \emptyset \land Y = X \cup \{0_C\}]\!]$$
$$\le [\![U^+ \cap X \ne \emptyset]\!],$$

and (b) follows. □

Remarks. (1) Taking $B = 2$ and $C = \hat{A}$ in 6.20, where A is a complete Boolean algebra in V, then $2 \otimes \hat{A} \cong A$ and it is not hard to see that U^+ is (essentially) the canonical generic ultrafilter in $V^{(A)}$. So 6.20 generalizes 4.21.

(2) Since $V^{(B)} \subseteq V^{(B \otimes C)}$, we may regard $V^{(B)}$ as a *class* in $V^{(B \otimes C)}$. Moreover, it is not hard to see that, within $V^{(B \otimes C)}$, $V^{(B)}$ is a transitive model of ZFC containing all the ordinals. So working inside $V^{(B \otimes C)}$ we can form the C-extension $(V^{(B)})^{(C)}$ of $V^{(B)}$. Now Theorem 6.20 may be construed as asserting that, within $V^{(B \otimes C)}$, U^+ is a $V^{(B)}$-generic ultrafilter in C. Accordingly within $V^{(B \otimes C)}$ we can form the transitive collapse $V^{(B)}[U^+]$ of the quotient $(V^{(B)})^{(C)}/U^+$ and by applying 4.22 within $V^{(B \otimes C)}$ we have

$$V^{(B \otimes C)} \models V^{(B)}[U^+] \text{ is the model of ZFC generated by } U^+ \text{ and } V^{(B)}.$$

Now, writing U_* for the canonical generic ultrafilter in $V^{(B \otimes C)}$, it is shown in Problem 6.36 that

$$V^{(B \otimes C)} \models U_* \in V^{(B)}[U^+]. \tag{6.21}$$

Moreover, we know from the remarks following 4.25 that

$$V^{(B \otimes C)} \models \forall x[x \in \hat{V}[U_*]] \tag{6.22}$$

6. Iterated Boolean Extensions

and so we get, using (6.21),

$$V^{(B \otimes C)} \models \forall x [x \in V^{(B)}[U^+]]. \tag{6.23}$$

In other words, $V^{(B \otimes C)}$ may be regarded as the Boolean-valued model of ZFC generated by $V^{(B)}$ and U^+.

Notice also that if U_o is the canonical generic ultrafilter in $V^{(B)}$ then $V^{(B)} \models \forall x[x \in \hat{V}[U_o]]$ and so we get, using (6.22) and (6.23),

$$V^{(B \otimes C)} \models \forall x [x \in \hat{V}[U_o][U^+]]$$
$$V^{(B \otimes C)} \models \hat{V}[U_*] = \hat{V}[U_o][U^+].$$

The first of these expressions tells us that $V^{(B \otimes C)}$ may be regarded as a "double generic extension" of $\hat{V}$, and the second that this double extension is expressible as a single extension.

(3) Working within $V^{(B \otimes C)}$, let $i_{U^+} := i$ be the canonical map of $(V^{(B)})^{(C)}$ onto $V^{(B)}[U^+]$. It follows from (6.23) that within $V^{(B \otimes C)}$ this map carries $(V^{(B)})^{(C)}$ onto the universe, i.e.

$$V^{(B \otimes C)} \models \forall x [x \in i[(V^{(B)})^{(C)}]]. \tag{6.24}$$

Now let $J^{(C)}$ be the class

$$J^{(C)} = \{x \in V^{(B)} : [\![x \in V^{(C)}]\!]^B = 1\};$$

that is, $J^{(C)}$ is the class of all B-valued sets which, with probability 1, are members of $(V^{(B)})^{(C)}$. $J^{(C)}$ may be deemed to be the class that represents $(V^{(B)})^{(C)}$ in the real world. We may regard $J^{(C)}$ as a $(B \otimes C)$-valued structure by defining, for $x, y \in J^{(C)}$,

$$[\![x = y]\!]^{J^{(C)}} = \text{unique } a \in B \otimes C \text{ such that } V^{(B)} \models a = [\![x = y]\!]^C$$
$$[\![x \in y]\!]^{J^{(C)}} = \text{unique } a \in B \otimes C \text{ such that } V^{(B)} \models a = [\![x \in y]\!]^C.$$

Next, let us agree to *identify* elements of $J^{(C)}$ (and $V^{(B \otimes C)}$) when they are equal with probability 1. Having done this, we can define the map $j: J^{(C)} \to V^{(B \otimes C)}$ by putting, for each $x \in J^{(C)}$,

$$j(x) = \text{unique } y \in V^{(B \otimes C)} \text{ such that } V^{(B \otimes C)} \models y = i(x).$$

Using (6.23) and (6.18) it is now easy to verify that j is an isomorphism of the Boolean-valued structures $J^{(C)}$ and $V^{(B \otimes C)}$. That is, j is a map of $J^{(C)}$ onto $V^{(B \otimes C)}$ such that $[\![x = y]\!]^{J^{(C)}} = [\![j(x) = j(y)]\!]^{B \otimes C}$ and $[\![x \in y]\!]^{J^{(C)}} = [\![j(x) \in j(y)]\!]^{B \otimes C}$ for all $x, y \in J^{(C)}$. Accordingly, we have shown that *the "iterated" Boolean extension $V^{(B)(C)}$ is equivalent to the "ordinary" Boolean extension $V^{(B \otimes C)}$*.

Our final result in this section is the

6.25. Lemma. *If B satisfies ccc and $V^{(B)} \models C$ satisfies ccc, then $B \otimes C$ satisfies ccc.*

Proof. Let $A = \{a_\xi : \xi < \omega_1\}$ be an antichain in $B \otimes C$; we show that there is $\xi_0 < \omega_1$ such that $a_\eta = 0$ for all $\eta > \xi_0$. Let $A' = A \times \{1\}$; then clearly

(1) $\quad V^{(B)} \models A'$ is an antichain in C.

Since C satisfies ccc in $V^{(B)}$, it follows from (1) that

(2) $\quad V^{(B)} \models A'$ is countable.

Now define $f \in V^{(B)}$ by

$$f = \{(\hat{\xi}, a_\xi)^{(B)} : \xi < \omega_1\} \times \{1\}.$$

It is then easily verified that

(3) $\quad V^{(B)} \models f : \hat{\omega}_1 \to A'$

and, for all $\xi < \omega_1$

(4) $\quad [\![f(\hat{\xi}) = a_\xi]\!] = 1$.

Since B satisfies ccc, we have, by 1.51,

$$V^{(B)} \models \hat{\omega}_1 \text{ is uncountable}$$

and it follows from (2) and (3) that

$$V^{(B)} \models \exists \xi < \omega_1 \forall \eta > \xi \forall \eta' > \xi [f(\eta) = f(\eta')].$$

Therefore, by the Maximum Principle, there is $\xi \in V^{(B)}$ such that

6. Iterated Boolean Extensions

(5) $\quad V^{(B)} \models \xi < \hat{\omega}_1 \wedge \forall \eta > \xi \forall \eta' > \xi [f(\eta) = f(\eta')]$.

Since B satisfies ccc, by 1.51(v) there is an ordinal $\xi_0 < \omega_1$ such that $[\![\xi < \hat{\xi}_0]\!] = 1$. It follows from (5) that

$$V^{(B)} \models \forall \eta > \hat{\xi}_0 [f(\eta) = f(\hat{\xi}_0)].$$

Hence $[\![f(\hat{\eta}) = f(\hat{\xi}_0)]\!] = 1$ whenever $\xi_0 < \eta < \omega_1$. Thus, by (4), $[\![a_\xi = a_\eta]\!] = 1$ whenever $\xi_0 < \eta < \omega_1$. But since A is an antichain we have, for $\eta > \xi_0$,

$$1 = [\![a_{\xi_0} = a_\eta]\!] \leq [\![a_\eta = a_{\xi_0} \wedge a_\eta]\!] = [\![a_\eta = 0]\!].$$

Therefore $a_\eta = 0$ in $B \otimes C$ for $\eta > \xi_0$, as claimed. □

Further Results on Boolean Algebras

In this section we give some technical results on Boolean algebras that we shall require for the proof of relative consistency of SH.

Our first result is a generalization of 2.10.

6.26. Lemma. *Let B be a complete Boolean algebra satisfying ccc, and let D be a dense subset of B. Then for each $b \in B$ there is a countable subset D_b of D such that $b = \bigvee D_b$. Moreover, $|B| \leq |D|^{\aleph_0}$.*

Proof. Using Zorn's lemma, let D_b be a maximal antichain in the set $\{x \in D : x \leq b\}$. Put $a = \bigvee D_b$; we claim that $a = b$. Clearly $a \leq b$. On the other hand, consider $b - a$. If it is non-zero, then there is $d \in D$ such that $0 \neq d \leq b - a$. But then d is disjoint from every member of D_b, contradicting the latter's maximality. It follows that $b - a = 0$, so $a = b$.

Thus $b = \bigvee D_b$; and since B satisfies ccc, D_b, as an antichain, must be countable.

Consequently each member of B is determined by a countable subset of D; since there are at most $|D|^{\aleph_0}$ of these, the claimed inequality follows. □

6.27. Corollary. *Suppose κ is a regular uncountable cardinal, B is a complete Boolean algebra satisfying ccc such that $|B| \leq \kappa$ and $C \in V^{(B)}$*

satisfies

$V^{(B)} \models C$ *is a complete Boolean algebra, satisfies*
ccc and has a dense subset of cardinality $< \hat{\kappa}$.

Then for some cardinal $\lambda < \kappa$, $|B \otimes C| \leq \kappa^\lambda$.

Proof. Let $Q \in V^{(B)}$ be such that

$$[\![Q \text{ is dense in } C \text{ and } |Q| < \hat{\kappa}]\!] = 1$$

and let $Q' = \{x \in B \otimes C : [\![x \in Q]\!] = 1\}$. Then, using 1.51(v), there is an ordinal $\alpha < \kappa$ such that $[\![|Q| < \hat{\alpha}]\!] = 1$. Putting $\lambda = |\alpha| < \kappa$ (or $\lambda = \omega$ if α is finite), we claim first that $|Q'| \leq \kappa^\lambda$. To see this, observe that since $[\![|Q| < \hat{\alpha}]\!] = 1$ the Maximum Principle yields an $f \in V^{(B)}$ such that $[\![f \text{ is map of } \hat{\alpha} \text{ onto } Q]\!] = 1$. Each $x \in Q'$ is then uniquely determined by the function $g_x : \alpha \to B$ defined by $g_x(\gamma) = [\![f(\hat{\gamma}) = x]\!]$. Since there are at most κ^λ such functions g_x, the claim follows.

We next claim that the set

$$S = \{b \wedge x : 0 \neq b \in B \wedge x \in Q'\}$$

is dense in $B \otimes C$. For suppose $0 \neq c \in B \otimes C$. Then $[\![c \in C]\!] = 1$ and $[\![c \neq 0_C]\!] = b \neq 0$. Put $d = b.c + b^*.1$. Then $[\![d = c]\!] \geq b$, $[\![d \in C]\!] = 1$ and $[\![d \neq 0_C]\!] = 1$. Since $[\![Q \text{ is dense in } C]\!] = 1$, there is $x \in Q'$ such that $[\![x \leq_C d]\!] = 1$. It follows that

$$b \leq [\![c = d]\!] = [\![c = d]\!] \wedge [\![x \leq_C d]\!] \leq [\![x \leq_C c]\!].$$

So, by (6.19), $b \wedge x \leq c$, and the claim follows.

Now we have $|S| \leq |Q'|.|B| \leq \kappa^\lambda . \kappa = \kappa^\lambda$. Therefore, by 6.26,

$$|B \otimes C| \leq |S|^{\aleph_0} \leq \kappa^{\lambda . \aleph_0} = \kappa^\lambda. \quad \square$$

We recall (see the remark after 2.3) that for each Boolean algebra A there is a complete Boolean algebra B called the *completion* of A such that (if we identify A as a subalgebra of B via the canonical monomorphism $f : A \to B$)

6. Iterated Boolean Extensions

(i) A is a subalgebra of B;

(ii) $A - \{0\}$ is dense in B;

(iii) if $X \subseteq A$ has a join $\bigvee_A X$ in A, then $\bigvee_A X = \bigvee_B X$.

Let $\langle B_i : i \in I \rangle$ be a chain of (complete) Boolean algebras; *i.e.* such that for any pair B_i, B_j, one is a subalgebra of the other. Then we can form the direct limit $\varinjlim B_i$ of the chain in the category of Boolean algebras in the usual way: $\varinjlim B_i$ is just $\bigcup_{i \in I} B_i$ with Boolean operations inherited from the B_i in the obvious manner. The completion of $\varinjlim B_i$ is called the *limit completion* of the chain $\langle B_i : i \in I \rangle$ and is written $\mathrm{limco}_{i \in I} B_i$ or $\mathrm{limco}\, B_i$. Clearly $\mathrm{limco}\, B_i$ includes $\bigcup_{i \in I} B_i - \{0\}$ as a dense subset.

Now let α be a limit ordinal. A sequence $\langle B_\xi : \xi < \alpha \rangle$ of complete Boolean algebras is called a *normal* sequence if

(a) $B_0 = 2$;

(b) for $\xi < \eta$, B_ξ is a complete subalgebra of B_η;

(c) for limit β, $B_\beta = \mathrm{limco}_{\xi < \beta} B_\xi$.

6.28. Lemma. *Let $\langle B_\xi : \xi < \alpha \rangle$ be a normal sequence of complete Boolean algebras and let $B = \mathrm{limco}_{\xi < \alpha} B_\xi$. Then each B_ξ is a complete subalgebra of B. Suppose, further, that each B_ξ satisfies ccc and that α is an uncountable regular cardinal. Then*

(i) $B = \bigcup_{\xi < \alpha} B_\xi$;

(ii) *if $|X| < \alpha$ and $f : X \to B$, then $\mathrm{ran}(f) \subseteq B_\xi$ for some $\xi < \alpha$.*

Proof. Let $X \subseteq B_\xi$ and $a = \bigvee X$ in B_ξ. We claim that $a = \bigvee X$ in $\varinjlim B_\eta = B'$. For if not, then X would have an upper bound $b < a$ in B'. Since $B' = \bigcup B_\eta$ there is $\eta < \alpha$ such that $b \in B_\eta$, and without loss of generality we may suppose that $\xi \leq \eta$. But then, in B_η, $\bigvee X \leq b < a$, contradicting the assumption that B_ξ is a complete subalgebra of B_η. Therefore $\bigvee_B X = \bigvee_{B'} X$ and, since B is the completion of B', $\bigvee_{B'} X =$

$\bigvee_B X$. So B_ξ is a complete subalgebra of B.

We next prove (i). Let $x \in B$. Since $\bigcup_{\xi < \alpha} B_\xi - \{0\}$ is dense in B, by 6.26 there is a countable set $\{x_n : n \in \omega\}$ such that $x_n \in B_{\xi_n}$ with $\xi_n < \alpha$ and $x = \bigvee_{n \in \omega} x_n$. Since λ is regular, there is $\xi < \alpha$ such that $\xi \geq \xi_n$ for all n, so that $x_n \in B_\xi$ for all n, whence $x = \bigvee_{n \in \omega} x_n \in B_\xi$. (i) follows.

Finally, (ii) follows easily from (i) and the regularity of α. □

6.29. Corollary. *Let κ be an uncountable regular cardinal, $\langle B_\xi : \xi < \kappa \rangle$ be a normal sequence of complete Boolean algebras satisfying ccc and let $B = \mathrm{limco}\, B_\xi$. Suppose further that $X \in V^{(B)}$ satisfies* (a) $[\![|X| < \kappa]\!] = 1$ *and* (b) *either* $[\![X \subseteq \hat{\kappa}]\!] = 1$ *or* $[\![X \subseteq \hat{\kappa} \times \hat{\kappa}]\!] = 1$ *or* $[\![X \subseteq P\hat{\kappa} \wedge |\bigcup X| < \hat{\kappa}]\!] = 1$. *Then there are $\xi < \kappa$ and $Y \in V^{(B_\xi)}$ such that $[\![Y = X]\!] = 1$.*

Proof. Since κ and $\kappa \times \kappa$ are naturally bijective and each $X \subseteq P\kappa$ such that $|X| < \kappa$ is naturally correlated with a subset of $\kappa \times \kappa$, the proof reduces to the case in which $[\![X \subseteq \hat{\kappa}]\!] = 1$. By 1.51(iv) we know that $[\![\hat{\kappa} \text{ is regular}]\!] = 1$ and since $[\![|X| < \hat{\kappa}]\!] = 1$ it follows (using the Maximum Principle) that there is $\alpha \in V^{(B)}$ such that

$$[\![\alpha < \hat{\kappa} \wedge \forall \xi \geq \alpha [\xi \notin X]]\!] = 1.$$

By 1.51(v) there is an ordinal $\gamma < \kappa$ such that $[\![\alpha < \hat{\gamma}]\!] = 1$. It follows that

(1) $\quad [\![\hat{\xi} \in X]\!] = 0 \quad$ for all $\xi \geq \gamma$.

Now define $Y \in V^{(B)}$ by $\mathrm{dom}(Y) = \{\hat{\xi} : \xi < \gamma\}$ and $Y(\hat{\xi}) = [\![\hat{\xi} \in X]\!]$. It is then easily verified, using (1), that $[\![X = Y]\!] = 1$. So it remains to show that $Y \in V^{(B_\xi)}$ for some $\xi < \kappa$. But this follows immediately from an application of 6.28 to the map $f : \gamma \to B$ given by $f(\xi) = Y(\hat{\xi})$. □

Now let B be a complete Boolean algebra and let C be a complete subalgebra of B. We define the map $\pi : B \to C$ as follows.

$$\pi(x) = \bigwedge \{y \in C : x \leq y\}.$$

Notice that we then have, for $b \in B$, $c \in C$, the *"adjointness"* condition:

$$b \leq c \leftrightarrow \pi(b) \leq c, \tag{6.30}$$

6. Iterated Boolean Extensions

which in turn characterizes π uniquely. The map π is called the *canonical projection* of B onto C.

6.31. Lemma. *The canonical projection $\pi : B \to C$ has the following properties:*

(i) $\pi(x) \geq x$;

(ii) $x \leq y \to \pi(x) \leq \pi(y)$;

(iii) $\pi(c) = c$ *for all* $c \in C$;

(iv) $\pi(\bigvee X) = \bigvee \pi[X]$ *for all* $X \subseteq B$;

(v) $\pi(b) \wedge c = \pi(b \wedge c)$ *for all* $b \in B, c \in C$.

Proof. (i), (ii), and (iii) are obvious. To prove (iv), we note that, if $c \in C$, then

$$\begin{aligned}
\pi(\bigvee X) \leq c &\leftrightarrow \bigvee X \leq c \quad &\text{(by (6.30))}\\
&\leftrightarrow \forall x \in X [x \leq c] \\
&\leftrightarrow \forall x \in X [\pi(x) \leq c] \quad &\text{(by (6.30))}\\
&\leftrightarrow \bigvee \pi[X] \leq c.
\end{aligned}$$

As for (v), we have $\pi(b) \wedge c = \pi(b) \wedge \pi(c) \geq \pi(b \wedge c)$ by (ii) and (iii). On the other hand,

$$\begin{aligned}
\pi(b) &= \pi(b \wedge c) \vee \pi(b \wedge c^*) \quad &\text{(by (iv))}\\
&\leq \pi(b \wedge c) \vee c^* \quad &\text{(by (ii) and (iii))}.
\end{aligned}$$

So $\pi(b) \wedge c \leq \pi(b \wedge c) \wedge c \leq \pi(b \wedge c)$ and the result follows. □

Our final preparatory result concerns the preservation of the countable chain condition under passage to limit completions.

6.32. Theorem. *Let $\langle B_\xi : \xi < \alpha \rangle$ be a normal sequence of complete Boolean algebras, and suppose that each B_ξ satisfies ccc. Then the limit completion B of $\langle B_\xi : \xi < \alpha \rangle$ also satisfies ccc.*

Proof. We first reduce the proof of the theorem to the case in which $\alpha = \omega_1$. Suppose the hypothesis of the theorem true and the conclusion false.

Let A be an antichain in B of cardinality $\aleph_1$ such that $0 \notin A$. Since $\bigcup_{\xi < \alpha} B_\xi$ is dense in B, for each $a \in A$ we can find an ordinal $\gamma_a < \alpha$ and an element $b_a \in B_{\gamma_a}$ such that $0 \neq b_a \leq a$. Clearly $\{b_a : a \in A\}$ is an antichain in B of cardinality $\aleph_1$. Also, $\sup\{\gamma_a : a \in A\} = \alpha$, for otherwise there would be an ordinal $\beta < \alpha$ for which $\gamma_a < \beta$ for $a \in A$, contradicting the assumption that B_β satisfies ccc. Since $|A| = \aleph_1$, it follows that α is cofinal with ω_1.

On the other hand, α is not cofinal with ω. For if $\alpha = \sup\{\alpha_n : n \in \omega\}$ then for each $a \in A$ there is $n \in \omega$ such that $\gamma_a < \alpha_n$. Hence $\{b_a : a \in A\} \subseteq \bigcup_{n \in \omega} B_{\alpha_n}$ and so, for some n, $B_{\alpha_n} \cap \{b_a : a \in A\}$ must have cardinality $\aleph_1$. This contradicts the assumption that B_{α_n} satisfies ccc.

It follows that α has cofinality ω_1. Accordingly there is a sequence of ordinals $\langle \beta_\xi : \xi < \omega_1 \rangle$ such that

$$\beta_0 = 0, \eta < \xi \to \beta_\eta < \beta_\xi, \alpha = \sup\{\beta_\xi : \xi < \omega_1\}$$
$$\text{and } \beta_\xi = \sup\{\beta_\eta : \eta < \xi\} \text{ for limit } \xi.$$

If we put B'_ξ for B_{β_ξ}, then $\langle B'_\xi : \xi < \omega_1 \rangle$ is a normal sequence, each B_ξ satisfies ccc, but the limit completion B_α of $\langle B'_\xi : \xi < \omega_1 \rangle$ does not.

Thus we need only prove the theorem when $\alpha = \omega_1$. Let X be an antichain in B; we show that X is countable.

Without loss of generality we may assume that $\bigvee X = 1$ and, since $\bigcup_{\xi < \omega_1} B_\xi - \{0\}$ is dense in B, that $X \subseteq \bigcup_{\xi < \omega_1} B_\xi$. For $\xi < \omega_1$, put $X_\xi = X \cap B_\xi$. Since B_ξ satisfies ccc, X_ξ is countable.

Now by 6.28, B_ξ is a complete subalgebra of B and so we can consider the canonical projection π_ξ of B onto B_ξ. By 6.31(iv), we have

$$1 = \pi_\xi(\bigvee X) = \bigvee \pi_\xi[X].$$

Since B_ξ satisfies ccc, there is, by 1.53(iv), a countable subset X' of X such that

$$1 = \bigvee \pi_\xi[X'].$$

It follows that for some countable ordinal δ_ξ we have $X' \subseteq X_{\delta_\xi}$ and

$$1 = \bigvee \pi_\xi[X_{\delta_\xi}]. \tag{1}$$

6. Iterated Boolean Extensions

Now let γ be a countable limit ordinal such that $\delta_\xi < \gamma$ for all $\xi < \gamma$. We claim that $X = X_\gamma$. Since X_γ is countable, this will complete the proof.

For $\xi < \gamma$ we have
$$\bigvee \pi_\xi[X_{\delta_\xi}] = \pi_\xi(\bigvee X_{\delta_\xi}) \leq \pi_\xi(\bigvee X_\gamma),$$
so it follows from (1) that
$$1 = \pi_\xi(\bigvee X_\gamma). \tag{2}$$

We claim that $\bigvee X_\gamma = 1$. For if not, then since $\bigcup_{\xi<\gamma} B_\xi - \{0\}$ is dense in B_γ, there is $\xi < \gamma$ and $0 \neq b \in B_\xi$ such that $b \wedge \bigvee X_\gamma = 0$. Hence, by 6.31(v)
$$0 = \pi_\xi(b \wedge \bigvee X_\gamma) = b \wedge \pi_\xi(\bigvee X_\gamma).$$
But this contradicts (2).

Therefore $\bigvee X_\gamma = 1$ and it follows easily from this that $X = X_\gamma$. □

Notice that 6.32 yields as an immediate corollary the result that the direct limit of a normal sequence of complete Boolean algebras satisfying ccc also satisfies ccc.

The Relative Consistency of SH

At long last we are in a position to prove the relative consistency of MA + $2^{\aleph_0} > \aleph_1$, and so of SH.

6.33. Theorem. *Let κ be an uncountable regular cardinal such that for any $0 \neq \lambda < \kappa$ we have $\kappa^\lambda = \kappa$. Then there is a complete Boolean algebra B such that*
$$V^{(B)} \models MA + 2^{\aleph_0} = \hat{\kappa}.$$

Proof. We construct recursively a normal sequence $\langle B_\xi : \xi < \kappa \rangle$ of complete Boolean algebras such that, for all $\xi < \kappa$, (a) B_ξ satisfies ccc and (b) $|B_\xi| \leq \kappa$.

For each complete Boolean algebra B satisfying ccc such that $|B| \leq \kappa$, let $D(B)$ be a core for the $V^{(B)}$-set $\{x : x \subseteq \hat{\kappa} \times \hat{\kappa} \wedge |x| < \hat{\kappa}\}^{(B)}$. We claim that $|D(B)| \leq \kappa$. To prove this we note first that if $x \in D(B)$ then, since $[\![\hat{\kappa} \text{ is regular }]\!] = 1$ by 1.51(iv) there is $\alpha \in V^{(B)}$ such that $[\![\alpha < \hat{\kappa} \wedge x \subseteq \alpha \times \alpha]\!] = 1$ and hence by 1.51(v) an ordinal $\beta < \kappa$ such that $[\![\alpha < \hat{\beta}]\!] = 1$. Therefore $[\![x \subseteq \hat{\beta} \times \hat{\beta}]\!] = 1$. It follows that if for each $\beta < \kappa$ we let v_β be a core for the $V^{(B)}$-set $P^{(B)}(\hat{\beta} \times \hat{\beta})$ then $|D(B)| \leq |\bigcup_{\beta<\kappa} v_\beta| \leq \Sigma_{\beta<\kappa} |v_\beta|$. Now if for each $z \in v_\beta$ we define $f_z : \beta \times \beta \to B$ by $f_z(\xi, \eta) = [\![\langle \hat{\xi}, \hat{\eta} \rangle \in z]\!]$, then it is easily verified that the map $z \mapsto f_z$ is a bijection between v_β and $B^{\beta \times \beta}$. It follows that $|v_\beta| \leq \kappa^{|\beta|} = \kappa$. Therefore $|D(B)| \leq \Sigma_{\beta<\kappa} |v_\beta| \leq \kappa.\kappa = \kappa$ as claimed.

For each complete Boolean algebra satisfying ccc such that $|B| \leq \kappa$, we fix an enumeration $\langle R^B_\xi : \xi < \kappa \rangle$ of $D(B)$.

Now we put $B_0 = 2$, and for α satisfying $0 < \alpha < \kappa$ assume as inductive hypothesis that $\langle B_\xi : \xi < \alpha \rangle$ has been constructed so as to be a normal sequence satisfying (a) and (b).

If α is a limit ordinal, we put $B_\alpha = \text{limco}_{\xi<\alpha} B_\xi$. If α is a successor ordinal, say $\xi+1$, we construct $B_\alpha = B_{\xi+1}$ as follows. Let $\xi \mapsto \langle \beta_\xi, \gamma_\xi \rangle$ be the canonical map of κ onto $\kappa \times \kappa$ (see Chapter 0). Recall that $\beta_\xi \leq \xi$ for any $\xi < \kappa$. Let $A = B_{\beta_\xi}$; then by inductive hypothesis A is a complete subalgebra of B_ξ and $|A| \leq \kappa$. Putting $R = R^A_{\gamma_\xi}$, we have $R \in V^{(B_\xi)}$. In $V^{(B_\xi)}$, let $\Delta(R) = \langle \{x : \langle x, x \rangle \in R\}, R \rangle$, and put

$$b = [\![\Delta(R) \text{ is a Boolean algebra satisfying ccc}]\!].$$

Let C'_ξ be the two-term mixture:

$$C'_\xi = b.\Delta(R) + b^*.\hat{2};$$

then

$$[\![C'_\xi \text{ is a Boolean algebra satisfying ccc}]\!] = 1.$$

Let $C_\xi \in V^{(B)}$ satisfy

$$V^{(B_\xi)} \models C_\xi \text{ is the completion of } C'_\xi.$$

6. Iterated Boolean Extensions

Note that
$$V^{(B_\xi)} \models C_\xi \text{ satisfies ccc}.$$

We finally put
$$B_{\xi+1} = B_\xi \otimes C_\xi,$$

and identify B_ξ with its canonical image in $B_\xi \otimes C_\xi$, so that B_ξ becomes a complete subalgebra of $B_{\xi+1}$.

Clearly $\langle B_\xi : \xi < \kappa \rangle$ constructed in this way is a normal sequence. We now verify (a) and (b).

(a) is proved by induction on ξ. $B_0 = 2$ clearly satisfies ccc. If B_ξ satisfies ccc, then $B_{\xi+1} = B_\xi \otimes C_\xi$ satisfies ccc by 6.25. If α is a limit ordinal and B_ξ satisfies ccc for all $\xi < \alpha$, then B_α, as the limit completion of the normal sequence $\langle B_\xi : \xi < \alpha \rangle$, satisfies ccc by 6.32. So (a) follows.

(b) is also proved by induction on ξ. Clearly $|B_0| \leq \kappa$. If α is a limit ordinal then B_α has the dense subset $\bigcup_{\xi < \alpha} B_\xi - \{0\}$ of cardinality $\leq \kappa$ by inductive hypothesis, so, since B_α satisfies ccc, $|B_\alpha| \leq \kappa^{\aleph_0} = \kappa$ by 6.26. If $\xi < \kappa$ and $|B_\xi| \leq \kappa$, we observe that

$$V^{(B_\xi)} \models C_\xi \text{ satisfies ccc and has a dense subset } C'_\xi \text{ such that } |C'_\xi| < \hat{\kappa},$$

so $|B_{\xi+1}| = |B_\xi \otimes C_\xi| \leq \kappa$ by 6.27.

Now we define B to be the limit completion of $\langle B_\xi : \xi < \kappa \rangle$. Then by 6.32 B satisfies ccc and since it has the dense subset $\bigcup_{\xi < \kappa} B_\xi - \{0\}$ of cardinality $\leq \kappa$, it follows from 6.26 that $|B| \leq \kappa^{\aleph_0} = \kappa$. Therefore by 2.19(i), we have

$$V^{(B)} \models 2^{\aleph_0} \leq (\kappa^{\aleph_0})\hat{\,} = \hat{\kappa}.$$

Thus, if we can show that $\text{MA}_{\hat{\kappa}}$ holds in $V^{(B)}$, we will have, by 6.6, $V^{(B)} \models \hat{\kappa} \leq 2^{\aleph_0}$, whence $V^{(B)} \models \hat{\kappa} = 2^{\aleph_0}$ and so, in $V^{(B)}$, $\text{MA}_{\hat{\kappa}}$ is just MA. It will therefore follow that

$$V^{(B)} \models 2^{\aleph_0} = \hat{\kappa} \wedge \text{MA}.$$

6. Iterated Boolean Extensions

So it remains to show that

$$V^{(B)} \models \mathrm{MA}_{\hat{\kappa}}.$$

By 6.8 and 1.28(ii) it suffices to show that, if $A, R, S \in V^{(B)}$ satisfy

(1) $V^{(B)} \models \langle A, R \rangle$ *is a Boolean algebra satisfying* $\mathrm{ccc} \wedge |A| < \hat{\kappa} \wedge S \subseteq PA \wedge |S| < \hat{\kappa}$,

then

$$V^{(B)} \models \text{there is an } S\text{-complete ultrafilter in } \langle A, R \rangle.$$

Without loss of generality we may assume that $V^{(B)} \models A \subseteq \hat{\kappa}$ and so

(2) $V^{(B)} \models R \subseteq \hat{\kappa} \times \hat{\kappa} \wedge |R| < \hat{\kappa} \wedge S \subseteq P\hat{\kappa} \wedge |S \cup \bigcup S| < \hat{\kappa}$.

By 6.29, there exist $\xi < \kappa$ and $A', R', S' \in V^{(B_\xi)}$ such that $[\![A = A']\!] = [\![R = R']\!] = [\![S = S']\!] = 1$ and so we may assume that $A, R, S \in V^{(B_\xi)}$.

Now we claim that

(3) $V^{(B_\xi)} \models \langle A, R \rangle$ *is a Boolean algebra satisfying* $\mathrm{ccc} \wedge |A| < \hat{\kappa} \wedge S \subseteq PA \wedge |S| < \hat{\kappa}$.

For suppose, e.g. $[\![|A| < \hat{\kappa}]\!]^{B_\xi} \neq 1$. Then $[\![|A| \geq \hat{\kappa}]\!]^{B_\xi} \neq 0$, so, for some $f \in V^{(B_\xi)}$, $[\![f : \hat{\kappa} \xrightarrow{\text{one-one}} A]\!]^{B_\xi} \neq 0$. By 1.21, and the fact that B_ξ is a complete subalgebra of B, $[\![f : \hat{\kappa} \xrightarrow{\text{one-one}} A]\!]^B \neq 0$. Therefore $[\![|\hat{\kappa}| \leq |A|]\!]^B \neq 0$. But since B satisfies ccc, we have $[\![\hat{\kappa} = |\hat{\kappa}|]\!]^B = 1$, so $[\![\hat{\kappa} \leq |A|]\!] \neq 0$. This contradicts (1).

Similarly, if

$$[\![\langle A, R \rangle \text{ does not satisfy ccc}]\!]^{B_\xi} \neq 0,$$

then there is $X \in V^{(B_\xi)}$ such that

$$[\![X \text{ is an antichain in } A \wedge |X| \geq \aleph_1 = \hat{\aleph}_1]\!]^{B_\xi} \neq 0.$$

So by 1.21

$$[\![X \text{ is an antichain in } A \wedge |X| \geq \aleph_1]\!]^B \neq 0,$$

whence

$$[\![X \text{ is an uncountable antichain in } A]\!]^B \neq 0.$$

But this contradicts (1). So (3) is proved.

It follows from (2) and the definition of $D(B_\xi)$ that there is $R'' \in D(B_\xi)$ such that $[\![R = R'']\!]^{B_\xi} = 1$ and so without loss of generality we may assume that $R \in D(B_\xi)$. Choose $\eta < \kappa$ to satisfy $R = R_\eta^{B_\xi}$, and $\alpha < \kappa$ so that

$$\beta_\alpha = \xi, \gamma_\alpha = \eta, \alpha \geq \beta_\alpha = \xi.$$

Then A, R and S are all in $V^{(B_\alpha)}$ and reasoning similar to that used to prove (3) shows that

$$V^{(B_\alpha)} \models \langle A, R \rangle \text{ is a Boolean algebra satisfying ccc }.$$

We claim that an S-complete ultrafilter in $\langle A, R \rangle$ was added in the passage from $V^{(B_\alpha)}$ to $V^{(B_{\alpha+1})}$.

Now $B_{\alpha+1} = B_\alpha \otimes C_\alpha$, so by 6.20 the object $U^+ \in V^{(B_{\alpha+1})}$ satisfies

$$V^{(B_{\alpha+1})} \models U^+ \text{ is a } P^{(B_\alpha)}(C_\alpha)\text{-complete ultrafilter in } C_\alpha.$$

But, by construction we have

$$V^{(B_\alpha)} \models C_\alpha \text{ is the completion of } \langle A, R \rangle$$

and since $V^{(B_{\alpha+1})} \models S \subseteq P^{(B_\alpha)}(C_\alpha)$ it follows that

$$V^{(B_{\alpha+1})} \models U^+ \cap A \text{ is an } S\text{-complete ultrafilter in } \langle A, R \rangle,$$

as required.

Since $B_{\alpha+1}$ is a complete subalgebra of B, we get

$$V^{(B)} \models U^+ \cap A \text{ is an } S\text{-complete ultrafilter in } \langle A, R \rangle,$$

and the proof is complete. $\square$

6.34. Corollary. *If ZF is consistent, so is* $\text{ZFC} + \text{SH}(+ \neg \text{CH})$.

Proof. Assuming GCH, take $\kappa = \aleph_2$ in 6.33. We get B such that $V^{(B)} \models$ MA $+ 2^{\aleph_0} = \aleph_2$, whence, by 6.9, $V^{(B)} \models$ SH. The result now follows from 1.19. □

We conclude this chapter with some remarks on further results about SH that have been obtained. Jensen has shown that $V = L \to \neg$ SH, which of course yields another proof of the independence of SH. For an account of this proof, see Devlin (1977). Jensen has also shown that SH + GCH is relatively consistent with ZFC, but the proof is very involved: see Devlin and Johnsbraten (1974).

For more applications of Martin's axiom, see Rudin (1977).

Problems

6.35. (*The iteration theorem*). Let M be a transitive $\in$-model of ZFC, let B be a complete Boolean algebra in the sense of M, let C, $\leq_C$ be elements of $M^{(B)}$ such that $M^{(B)} \models \langle C, \leq_C \rangle$ *is a complete Boolean algebra*, and let $D = B \otimes C$. We identify B as a complete subalgebra of D.

(i) Let F be an M-generic ultrafilter in B and let i_F be the canonical map of $M^{(B)}$ onto $M[F]$. Then $i_F(C)$ is a complete Boolean algebra (with partial ordering $i_F(\leq_C)$) in $M[F]$. Show that $i_F \mid D$ is an M-complete homomorphism of D onto $i_F(C)$, *i.e.* preserves the join of any subset of D which is at the same time a member of M. (Clearly $i_F[D] \subseteq i_F(C)$. To prove equality, use 1.32. To show that i_F is M-complete, use the M-genericity of F in the form: $A \subseteq F, A \in M \to \bigwedge A \in F$.)

(ii) *A double generic extension is equivalent to a single one.* Let F be an M-generic ultrafilter in B, and let G be an $M[F]$-generic ultrafilter in $i_F(C)$. Put $H = i_F^{-1}[G] \cap D$. Show that H is an M-generic ultrafilter in D and that $M[H] = M[F][G]$. (Use (i).)

(iii) . Let H be an M-generic ultrafilter in D. Then $F = B \cap H$ is an M-generic ultrafilter in B. Put $G = i_F[H]$; show that G is an $M[F]$-generic ultrafilter in $i_F(C)$ and that $M[H] = M[F][G]$. (Use 6.20 and the remarks following it.)

6. Iterated Boolean Extensions

6.36. (*More on* $\otimes$). Let B be a complete Boolean algebra and suppose that $V^{(B)} \models \langle \check{C}, \leq_C \rangle$ *is a Boolean algebra.* Put $A = B \otimes C$ and identify B as a subalgebra of A (see 6.15). Then $V^{(B)} \models \hat{A}$ and $\hat{B}$ *are Boolean algebras and* $\hat{B}$ *is a subalgebra of* $\hat{A}$.

(i) Show that, for $a \in A$, we have $[\![a = 1_C]\!] = \bigvee \{b \in B : b \leq a\}$. (Use (6.19).)

(ii) Define $h \in V^{(B)}$ by $h = \{\langle \hat{a}, a \rangle^{(B)} : a \in A\} \times \{1\}$. Show that $V^{(B)} \models h$ *is a homomorphism of* $\hat{A}$ *onto* C. (Note that $[\![h(\hat{a}) = a]\!] = 1$ for $a \in A$.)

(iii) Let U_0 be the canonical generic ultrafilter in $\hat{B}$ and, in $V^{(B)}$, let $F = \{x \in \hat{A} : \exists y \in U_0 [y \leq x]\}^{(B)}$ be the filter in $\hat{A}$ generated by U_0. Show that $V^{(B)} \models \forall x \in A[h(x) = 1 \leftrightarrow x \in F]$ and deduce that $V^{(B)} \models C \cong \hat{A}/F$. (Use (i).)

(iv) Let $U^+ \in V^{(A)}$ be the ultrafilter in C defined just before (6.18) (cf. 6.20) and let $U_* \in V^{(A)}$ be the canonical generic ultrafilter in $\hat{A}$. Show that $V^{(A)} \models h^{-1}[U^+] = U_*$ and deduce that $V^{(A)} \models U_* \in V^{(B)}[U^+]$.

6.37. (*The operation inverse to* $\otimes$). Let A be a complete Boolean algebra and let B be a complete subalgebra of A. Working inside $V^{(B)}$, let F be the filter in $\hat{A}$ generated by the canonical generic ultrafilter in $\hat{B}$ (cf. 6.35(iii)), let $A * B = \hat{A}/F$ and let $\pi : \hat{A} \to A * B$ be the natural epimorphism. Finally, let $p : A \to V^{(B)}$ be defined by $[\![p(a) = \pi(\hat{a})]\!] = 1$ for $a \in A$.

(i) Show that, for $x, y \in A$, $b \in B$, $b \wedge x \leq b \wedge y \leftrightarrow b \leq [\![p(x) \leq p(y)]\!]$ and $b \wedge x = b \wedge y \leftrightarrow b \leq [\![p(x) = p(y)]\!]$. (Like 6.16.)

(ii) Let $t \in V^{(B)}$ satisfy $[\![t \in A * B]\!] = 1$. Show that $[\![t = p(a)]\!] = 1$ for some $a \in A$. (Take a to be a suitable mixture of members of A, and use (i).)

(iii) Show that $[\![A * B \text{ is complete }]\!] = 1$. (Given $X \in V^{(B)}$ such that $[\![X \subseteq A * B \wedge 0 \in X]\!] = 1$, let $a = \bigvee \{x \in A : [\![p(x) \in X]\!] = 1\}$; use (i) and (ii) to show that $[\![a = \bigvee X]\!] = 1$.)

(iv) Show that $A \cong B \otimes (A * B)$. (Show that p does the trick.)

(v) Show, inversely, that if

$$V^{(B)} \models C \text{ is a complete Boolean algebra},$$

then $C \cong (B \otimes C) * B$. (Use 6.36(iii).)

6.38. (*Injective Boolean algebras*). A Boolean algebra B is said to be *injective* if, for any Boolean algebra A, any homomorphism of a subalgebra of A into B can be extended to the whole of A. B is said to be an *absolute subretract* if whenever B is a subalgebra of a Boolean algebra A there is an epimorphism from A to B which is the identity on B.

(i) Let B be a complete Boolean algebra and let U_* be the canonical generic ultrafilter in $V^{(B)}$. Show that the following conditions are equivalent:

(a) B is injective;

(b) B is an absolute subretract;

(c) for any Boolean algebra A of which B is a subalgebra, there is $U \in V^{(B)}$ such that $V^{(B)} \models U$ is an ultrafilter in (the Boolean algebra) $\hat{A}$ and $U_* \subseteq U$;

(d) for any $C \in V^{(B)}$ such that $V^{(B)} \models C$ is a Boolean algebra, there is $U \in V^{(B)}$ such that $V^{(B)} \models U$ is an ultrafilter in C. (For (iii) $\to$ (iv), use 6.36(iii).)

(ii) Deduce the *Sikorski Extension Theorem*: any complete Boolean algebra is injective.

Historical Notes

Chapter 1. Boolean-valued models of set theory were first introduced by Scott and Solovay (see Scott 1967) and Vopěnka (1967). Most of the results of this chapter appear in Scott (1967).

Chapter 2. The method of forcing was invented by Cohen (1963), (1964), who also proved the independence of the axiom of constructibility and of the continuum hypothesis. Corollary 2.13 is due to Solovay (1965). The Boolean valued versions of these results appear in Scott (1967) and Vopěnka (1967). For an approach to forcing closely related to the Boolean-valued one given here, see Shoenfield (1971).

The results in Problems 2.14–2.16 appear in Scott (1967). Problem 2.20 is due to Solovay (1963).

Chapter 3. The independence of the axiom of choice from ZF was established by Cohen in 1963 (see Cohen 1966). The Boolean-valued version may be found in Scott (1967) and Vopěnka (1967). Corollary 3.9 is due to Feferman (1965).

Chapter 4. The construction of generic extensions of models of set theory is due to Cohen (1963) (1964); the approach in this chapter is closely related to that of Shoenfield (1971). Theorems 4.6, 4.15 and 4.19 are from Bell (1976 a). The result in Problem 4.27 is due to Mansfield and Dawson (1976), that in 4.33 to Bell (1976), 4.36 to Solovay (1970), 4.37 to Solovay (cf. also Grigorieff (1975)), 4.38 and 4.39 to Vopěnka (cf. also Grigorieff 1975).

Chapter 5. The concept of a collapsing algebra, Corollary 5.2, Problem 5.3 and Problem 5.4 are due to Levy. The result in Problem 5.5 is due to

Levy and Solovay (1967). Theorem 5.6 is a result of Solovay (1966) and Theorem 5.7 of Kripke (1967). Problem 5.9 is due to Bell (1975).

Chapter 6. The formulation of SH in terms of trees is due to Miller (1943). The independence of Souslin's hypothesis from ZFC is due independently to Jech (1967) and Tennenbaum (1968): I have elected to present the latter's construction. The relative consistency of Souslin's hypothesis is due to Solovay and Tennenbaum (1971): my exposition is largely based on this paper. Martin's axiom (which was independently formulated by Rowbottom) makes its first appearance in print in Martin and Solovay (1970) where various applications are presented.

Bibliography

Bell, J.L. (1975). *A characterization of universal complete Boolean algebras.* J. Lond. Math. Soc. (2) **12**, 86–88.

Bell, J.L. (1976). *Uncountable standard models of* $ZFC + V \neq L$. Set Theory and Hierarchy Theory: A Memorial Tribute to A. Mostowski, Bierutowice, Poland 1975. Lecture Notes in Mathematics 537, Springer, Berlin-Heidelberg-New York.

Bell, J.L. (1976a.). *A note on generic ultrafilters.* Z. Math. Logik **22**, 307–310.

Bell, J.L. and Machover, M. (1977). *A course in mathematical logic.* North-Holland, Amsterdam.

Cohen, P.J. (1963). *The independence of the continuum hypothesis I.* Proc. Natn. Acad. Sci. U.S.A. **50**, 1143–1148.

Cohen, P.J. (1964). *The independence of the continuum hypothesis II.* Proc. Natn. Acad. Sci. U.S.A. **51**, 105–110.

Cohen, P.J. (1966). *Set theory and the continuum hypothesis.* Benjamin, New York.

Devlin, K.J. (1977). *Constructibility.* In: J. Barwise, ed. *Handbook of mathematical logic.* North-Holland, Amsterdam.

Devlin, K.J. and Johnsbraten, H. (1974). *The Souslin problem.* Lecture Notes in Mathematics 405, Springer, Berlin-Heidelberg-New York.

BIBLIOGRAPHY

Drake, F.R. (1974). *Set theory: an introduction to large cardinals.* North-Holland, Amsterdam.

Easton, W.B. (1970). *Powers of regular cardinals.* Ann. Math. Logic **1**, 141–178.

Feferman, S. (1965). *Some applications of the notions of forcing and generic sets.* Fund. Math. **56**, 325–345.

Felgner, U. (1971). *Models of ZF-set theory.* Lecture Notes in Mathematics 223, Springer, Berlin-Heidelberg-New York.

Fitting, M.C. (1969). *Intuitionistic logic, model theory and forcing.* North-Holland, Amsterdam.

Grigorieff, S. (1975). *Intermediate submodels and generic extensions in set theory.* Ann. Math. **101**, 447–490.

Halmos, P.R. (1963). *Lectures on Boolean algebras.* Van Nostrand, New York.

Jech, T.J. (1967). *Non-provability of Souslin's hypothesis.* Comment. Math. Universitatis Carolinae **8**, 291–305.

Jech, T.J. (1971). *Lectures in set theory.* Lecture Notes in Mathematics 217, Springer, Berlin-Heidelberg-New York.

Jech, T.J. (1973). *The axiom of choice.* North-Holland, Amsterdam.

Jech, T.J., ed. (1974). *Axiomatic set theory.* AMS Proceedings of Symposia in Pure Mathematics, Vol. XIII, Part II. American Mathematical Society, Providence.

Kelley, J.L. (1955). *General topology.* Van Nostrand, New York.

Kripke, S. (1967). *An extension of a theorem of Gaifman-Hales-Solovay.* Fund. Math. **61**, 29–32.

Levy, A. (1965). *Definability in axiomatic set theory I.* Proceedings of the 1964 International Congress on Logic, Methodology and Philosophy of Science. North-Holland, Amsterdam.

Levy, A. and Solovay, R.M. (1967). *Measurable cardinals and the continuum hyppothesis.* Israel J. Math. **5**, 234–248.

Mansfield, R. and Dawson, J. (1976). *Boolean-valued set theory and forcing.* Synthese **33**, 223–252.

Martin, D.A., and Solovay, R.M. (1970). *Internal Cohen extensions.* Ann. Math. Logic **2**, 143–178.

Mathias, A.R.D. *Lectures on Boolean-valued models for set theory.* (Unpublished notes.)

Miller, E.W. (1943). *A note on Souslin's problem.* Amer. Jour. Math. **65**, 673–678.

Rasiowa, H. and Sikorski, R. (1963). *The mathematics of metamathematics.* PWN, Warsaw.

Rosser, J.B. (1969). *Simplified independence proofs: Boolean-valued models of set theory.* Academic Press, New York.

Rudin, M.E. (1977). *Martin's axiom.* In J. Barwise, ed. *Handbook of mathematical logic.* North-Holland, Amsterdam.

Scott, D.S. (1967). *Boolean-valued models for set theory.* Mimeographed notes for the 1967 American Math. Soc. Symposium on axiomatic set theory.

Scott, D.S., ed. (1971). *Axiomatic set theory.* AMS Proceedings of Symposia in Pure Mathematics, Vol. XIII, Part I. American Mathematical Society, Providence.

Shoenfield, J.R. (1971). *Unramified forcing.* In Scott (1971).

Sikorski, R. (1964). *Boolean algebras.* Springer, Berlin-Heidelberg-New York.

Solovay, R.M. (1965). $2^{\aleph_0}$ *can be anything it ought to be.* In Addison, J.W., Henkin, L. and Tarski, A., eds. *The theory of models.* North-Holland, Amsterdam.

Solovay, R.M. (1966). *New proof of a theorem of Gaifman and Hales.* Bull. Am. Math. Soc. **72**, 282–284.

Solovay, R.M. (1970). *A model of set theory in which every set of reals is Lebesgue measurable.* Ann. Math. **92**, 1–56.

Solovay, R.M. and Tennenbaum, S. (1971). *Iterated Cohen extensions and Souslin's problem.* Ann. Math. **94**, 201–245.

Suzuki, Y. *Boolean-valued models for set theory.* (Unpublished notes.)

Takeuti, G. and Zaring, W.M. (1973). *Axiomatic set theory.* Springer, Berlin-Heidelberg-New York.

Tennenbaum, S. (1968). *Souslin's problem.* Proc. Nat. Acad. Sci. U.S.A. **59**, 60–63.

Vopěnka, P. (1965). *The limits of sheaves and applications on constructions of models.* Bull. Acad. Polon. Sci. Ser. Sci. Math. Astron. Phys. **13**, 189–192.

Vopěnka, P. (1967). *General theory of ∇-models.* Commentat. Math. Univ. Carolinae **8**, 145–170.

Vopěnka, P. and Hajek, P. (1972). *The theory of semisets.* North-Holland, Amsterdam.

Index of Symbols

$A * B$ 153
AC 7
$a_0.u_0 + a_1.u_1$ 25
$\text{Aut}(B)$ 67
$B \otimes C$ 132
B^+ 111
$C_\kappa(x,y)$ 64
$C(x,y)$ 51
ccc 42, 124
ccg 117
cf 63
$\text{Comp}(p,q)$ 48
$\text{Consis}(T)$ 19
$\text{dom}(f)$ 5
$f[X]$ 5
$f \mid x$ 5
$\text{Fun}(f)$ 5
G_J 81
G_n 81
G_* 106
G_{**} 107
$g.x$ 67
gx 67
$\bar{h}$ 73
$\tilde{h}$ 74
HOD 73
$(\text{HODM})^{M[U]}$ 111
i 91
i_U 91
j 91

$J^{(C)}$ 139
κ^+ 7
κ-cc 46
κ^λ 7
L 5
$\mathcal{L}$ 5
L_α 39
$\text{L}(x)$ 5
$\mathcal{L}^{(B)}$ 12
$\mathcal{L}_M^{(B)}$ 87
$\mathcal{L}^{(\Gamma)}$ 78
$\mathcal{L}_S$ 111
$\mathcal{L}_S^{(B)}$ 111
$\lim_{\rightarrow} B_i$ 143
$\text{limco}_{i \in I} B_i$ 143
M 87
$M^{(B)}$ 87
$M^{(B)}/U$ 88
$M[G]$ 105
$M[U]$ 91
MA 130
MA_κ 128
$\text{MA}_\kappa!$ 129
$N(p)$ 51
$(\text{ODM})^{M[U]}$ 111
O_p 48
$o(x)$ 122
OD 73
ORD 5
$\text{ORD}^{(M)}$ 89

INDEX OF SYMBOLS

$\mathrm{Ord}(x)$ 5
$P^{(B)}(u)$ 35
$P(x)$ 5
PX 3
Px 5
$\pi.b$ 67
$\mathrm{ran}(f)$ 5
$\mathrm{rank}(x)$ 8
$\mathrm{RO}(P)$ 48
$\mathrm{RO}(X)$ 3
S_1 89
S_2 92
S_3 94
SH 121
Σ_1 9
stab 77
$\sum_{i \in I} a_i.u_i$ 25
$t^{(M)}$ 9
U 88
U^+ 136
U_* 93
V 5
$\hat{V}$ 100
V_α 8
$V^{(B)}$ 11
$V^{(B)}_\alpha$ 11
$V^{(\Gamma)}$ 77
$V^{(\Gamma)}_\alpha$ 77
$V^{(2)}$ 10
$V^{(2)}_\alpha$ 10
WO_ϕ 71
$\hat{x}$ 22, 88
x^U 88
x^y 5
ZF 6
ZFC 7

ZFM 116
0 1
0_B 1
1 1
1_B 1
2 3
$|x|$ 7
$\langle u, v \rangle^{(B)}$ 45
$\langle x, y \rangle$ 5
$[\![\cdot]\!]^B$ 12
$[\![\sigma]\!]$ 87
$[\![\sigma]\!]^B$ 14
$[\![\sigma]\!]^\Gamma$ 78
$\{u, v\}^{(B)}$ 45
$\{u\}^{(B)}$ 45
$\{x \in u : \psi(x)\}^{(B)}$ 73
$\in$ 5
$\in_U$ 88
$\wedge$ 1
$\bigwedge$ 2
$\bigwedge_B$ 2
$\vee$ 1
$\bigvee$ 2
$\bigvee_B$ 2
$\sim_U$ 88
$\Rightarrow$ 2
$\Leftrightarrow$ 2
$*$ 1
$-$ 2
$\models$ 9
$\Vdash$ 52
$\Vdash_c$ 54
$\Vdash_\Gamma$ 79
$\leq$ 1
$\leq_{B \otimes C}$ 132

Index of Terms

action of G by automorphisms 67
antichain 25
automorphism 4
Axiom, Extensionality 6
—, Infinity 6
—, Power set 6
—, Regularity 6
—, Replacement 6
—, Separation 6
—, Union 6
— of Choice 7
— — Constructibility 7
B-extension of V 12
B-formula 12
B-sentence 12
basis 50
Boolean algebra 1
— completion 49, 51
— extension 101
— — of V 12
— valued model of ZFC 30
branch 122
canonical bijection 7
— generic set 107
— — ultrafilter 97, 100
— map of $M^{(B)}$ onto $M[U]$ 91
— projection 145
cardinal 7
— collapsing 113
characteristic function 10

Church's scheme 5
class of all ordinals 6
cofinality 63
collapsing $(\aleph_0, \lambda)$-algebra 115
— (κ, λ)-algebra 115
compatible 48
complete 2
— homomorphism 4
— subalgebra 20
completely embedded 119
completion of a Boolean algebra 50
continuum hypothesis 7
core 29
countable chain condition 42, 58, 124
countably M-complete 102
— completely generated 117
definable class 5
definite 35
dense 48, 105
— below an element 105
direct limit 143
extensional 18
Extensionality 6
F-complete 95
filter 4
— of subgroups 76
forces 52
formula 12

INDEX OF TERMS

free 122
Γ-forces 79
generalized continuum hypothesis 7
generic extension 97
group actions 67
— of automorphisms 4
height 122
hereditarily ordinal definable 73
holds with probability 1   15
homogeneous 70
homomorphism 4
Induction Principle for $V^{(B)}$ 12
Infinity 6
interpretation 9
invariant 70
involution 110
Iteration Lemma 106
join 2
κ-chain condition 46
κ-complete 115
κ-universal 120
(κ, λ)-distributive 61
$(\mathcal{L}\text{-})$structure 8
language of set theory 5
limit completion 143
M-generic 89, 105
— ultrafilter generated by G 105
M-partition of unity in B 89
Martin's axiom 130
— — at level κ 128
Maximum Principle 27
measurable 116
meet 2
Mixing Lemma 25
mixture 25

model of ZFC generated by U and M 98
— — — obtained by adjoining U to M 98
Mostowski's Collapsing Lemma 9
natural partial ordering 2
non-principal 115
normal filter 77
— sequence 143
order topology 48
ordered pair 45
ordinal definable 73
— —, hereditarily 73
— in $M^{(B)}/U$ 89
partition of unity 25
Power set 6
— — algebra 3
— — of u in $V^{(B)}$ 35
Product Lemma 106
Rasiowa-Sikorski Theorem 4
refined 48
— associate 51
regular 7
— open 3
— — algebra 3
Regularity 6
Replacement 6
restricted 9
— $(\kappa, 2)$-distributive law 62
S-complete 4
S-generic 129
sentence 12
Separation 6
set of conditions 50
Sikorski Extension Theorem 154
Souslin tree 122

Souslin's Hypothesis 121
standard 22
standard ordinal 38
strong forcing 54
subalgebra 3
transitive $\in$-model 9
— $\in$-structure 8
tree 122
true 15, 79
ultrafilter 4
Ultrafilter Theorem 4
ultrapower 75

Union 6
universe of 2-valued sets 11
— — constructible sets 6
— — sets 6
— $V^{(B)}$ of B-valued sets 11
valid 15
weak forcing 54
weakly (ω, κ)-distributive 62
well-founded relations 7
Zermelo-Fraenkel set theory 6
Zorn's Lemma 36
2-term mixture 25